**For Anja**

Calwius

# Nature Does Not Answer

## A Critique of Pure Science

tredition



Printing and distribution on behalf of the author:
tredition GmbH, Heinz-Beusen-Stieg 5, 22926 Ahrensburg, Germany

ISBN
Hardcover 978-3-384-03498-4
E-book 978-3-384-03499-1

# Contents

# Foreword

This book grew out of my unease about the influence of science on our lives. Whereas in the Middle Ages the church usually had the last word when it came to determining what was right and wrong, nowadays science has taken over this role.

The *critique of pure science* examines the so-called exact sciences, whose findings are scientific facts that differ from mere opinions, beliefs, and convictions. The focus is not on scientific action that attempts to solve problems, but exclusively on the scientific findings of the so-called basic sciences.

This book attempts to show that science's aspiration to explain to us why the world is the way it is, or why it appears to us the way it does, has failed.

# Introduction

There has been great technological progress since the Middle Ages. We now have a huge inventory of tools with which to impact very precisely on nature and to cause reactions of nature, whereby *tool* is to be understood in a very general way. A tool can be a machine, a device, or even a chemical compound. We do something and we know how nature reacts to it. We call this knowledge fact-based knowledge, because it is an indisputable fact that we know very precisely what our tools do. Because of this enormous increase in knowledge, we are led to assume that we completely understand nature or even the world because we are in possession of more and more knowledge about nature's response to more and more new tools.

But have we understood what nature is? Do we know the essence of nature?

Why nature reacts the way it does, we do not know. We can only speculate and make assumptions.

And this method only works really well when applied to inanimate nature. If we apply it to animate nature, the prediction of the reaction to an interaction is much less precise. There is also knowledge about reactions of living beings to interactions without using tools, but it is much less precise and therefore has a very low value.

Peter Sloterdijk put it very nicely into words:

> With the aid of Buddhist, Taoist, and original Christian, Indian, and American Indian intelligence, no production lines and no satellites can be built. However, in the modern type of knowledge, that awareness of life dries up from which the old teachings of wisdom take their inspiration, in order to speak of life and death, love and hate, antagonism and unity, individuality and cosmos, manliness and

womanliness. One of the most important motifs in the literature of wisdom is a warning against false cleverness, against "head" knowledge and learnedness, against thinking in terms of power and arrogant intellectuality.

Why we value statements about nature's reaction to our actions with the help of tools so highly will be examined in this book.

When we read texts by philosophers, the natural sciences are very often spoken of as exact sciences. In doing so, we feel a respect or even an admiration for this exactness. But what makes physics an exact science? In what way is physics more exact than philosophy?

Physics differs from philosophy in that it does not attempt to justify theses about the world through reasonable arguments. Instead, it makes experiments, and the measurement results of these experiments are regarded as facts and thus as evidence of whether these theses are right or wrong. Kant says of science that it is only the conceit of addressing the questions posed by reason to nature and not simply observing it that makes empiricism a science.

But how do we ask questions to nature? And how does nature answer?

According to physics, experiments, and measurements are the questions and measurement results are the answers. Behind this is the idea that all processes in the world can be explained by mathematical laws; that mathematical laws are the reason why all processes in the world are the way they are, and that measurement results help us to find these laws.

But why does physics believe that the world can be explained by mathematical laws? The reason lies in physics' idea that the whole world is a machine.

Physics begins with Galileo Galilei and reaches a pinnacle with Newton, and for Galilei and Newton the world was a machine created by God in which everything is fixed. It was only much later that God was dispensed with as creator, but the machine conception has remained.

An ideal machine does exactly what it was created to do. And since God created the world as a machine, the world is of course an ideal machine. And if we want to understand how a machine works, we look at the interplay of its parts, understand what mechanisms can be discerned, and gradually comprehend the whole machine and ultimately the purpose of that machine. So we can understand physics in the sense that God has created a machine whose function and purpose we do not know, but which we can find out by gradually looking at and understanding all its parts and their mechanical interaction.

If the world is a machine, then only the world as a whole has a goal and a purpose. Each part of a machine serves only the function of the machine and obeys only the mechanical laws that describe the functioning of the parts so that the machine functions. Everything that happens in the world is determined by the creation of the world as a machine that follows its mechanical laws.

It is a brilliant move of physics to reduce the task of understanding the world to finding laws - simply by believing that the whole world is a machine and obeys the laws of a machine. Whereas in the church of the Middle Ages, when people believed in the existence of God, they still had to find reasonable explanations why the world is the way it is, why God set this or that up this way, physics can do without this altogether. It is enough to find mathematical laws to understand the world, because God created the world as a machine and we only have to find the mechanism and the functioning of this machine to explain the world.

So naturally the question arises why physics believes that the world is a machine. Why does physics believe that goals and purposes play no role in describing phenomena and processes in the world? That everything is predetermined by mathematical laws? After all, our living world is by no means such that we have the impression that everything is predetermined. When we get up in the morning, we can spontaneously decide to put apricot jam on our toast instead of strawberry jam. In general, we make decisions throughout the day that do not seem predetermined in any way, and we are constantly surprised that things turn out differently than we thought. How can the idea arise in such a world that everything is in fact predetermined?

Physics manages this by a simple trick. It splits the world into *objective* and *subjective* and claims that all phenomena that do not fit into the machine model are subjective and do not really exist. For in the machine called the *world,* there are no spontaneous decisions. All perceptions and processes in the world are either to be explained as mechanical parts of a world machine, by finding its mathematical laws, or physics explains to us that these perceptions in reality – *objectively* – do not exist at all. If we do not find mathematical laws for a perception, then we only imagine – *subjectively* – that this perception really exists. Only when we find a mathematical law can we say that this perception really exists.

***

The first part of the book deals with the method of physics. Experiment and measurement are examined and it is shown that the method of physics does not ask nature any questions at all and that nature does not answer either.

In the second part, the concepts and ideas of physics are subjected to a philosophical examination.

The third part is about the role of science in society. Why physics could become the leading science and gradually replace the church as an authority. Why its method was adopted by other sciences and what actually happens when people engage in scientific activity.

The final chapter attempts an outlook on a different approach to science in society.

In the following chapters, various terms and ideas of physics are dealt with, and since all terms and ideas are interrelated, there is inevitably some repetition, but this is intentional for better understanding.

Notes, quotations, and references are mainly not included in the running text, but in a separate chapter at the end of the book.

# 1 Physics' method of understanding the world

While philosophy traditionally tried to understand why the world is the way it is, or why it appears to us the way it does, physics has a completely different concern. Instead of finding rational explanations why a stone falls to the ground when we let go of it, physics, for the first time in human history, asks how exactly the stone falls. The question of *why* is replaced by the question of *how*.

And the answer to the question "*How* does the stone fall?" is not answered by thinking, but by observing and measuring.

But why does physics believe it can do without the question of *why*? Why should it no longer matter *why* the stone falls? Because it replaces the *why* with mathematical laws. The mathematical law is the substitute for the rational explanation, because if there are indeed mathematical laws that describe the movement of objects, then the movement of these objects is fixed for all time. The past and the future of the movement are exactly described by the mathematical law. The reason why a stone is here now and there later is described by the mathematical law. We do not need any other reasonable explanations.

By describing movement mathematically, physics can dispense with the goal and purpose of a movement. Everything we observe is determined by the past and determines the future. If we go to Italy for a holiday, the goal of our movement is Italy and the purpose is the holiday there. For the stone that falls, physics dispenses with this description. The stone does not fall because it has the purpose of lying on the ground, and the stone also does not associate a purpose with lying on the ground. The stone falls because it obeys the mathematical law. Of course, the stone does not obey the law in the same way that we obey a law. In fact, this sentence is only a shortened form of saying that the law describes its movement and the construction of the *world* machine forces it to obey this law.

The task of physics is now to find out how fast or slow the stone falls and whether its speed is constant or whether its velocity changes.

To be able to solve this task, we need the possibility to measure velocities, and the measurement result must be a number. This number is the measure of speed. Because in order to be able to say that this stone falls twice or three times as fast as another stone, we need a measure of speed.

But physics wants even more. It is firmly convinced that there is an order in nature and that it can find out the rules of this order. That it is therefore no coincidence why one stone falls faster than another. That the height from which the stone falls has a direct influence on the speed with which it hits the ground. And physics is also absolutely convinced that the rules of this natural order are mathematical in nature. That we can therefore find mathematical laws that can accurately describe the falling of stones from different heights. Whether there is a mathematical law is not questioned at all. The question of physics is exclusively: which law applies?

And physics seeks this mathematical law not only for falling objects and for all objects that move, but also for all perceptions and their changes. The *why* of philosophy is replaced by the mathematical law. If physics finds a mathematical law that describes a phenomenon, then that is the reason for the observed phenomenon. So if we want to understand the world, physics tells us to find mathematical laws that describe the observed phenomena. That's all it takes. The mathematical law is the reason why the world is the way it is, or why it appears to us the way it appears to us.

To find a mathematical law, we have to make many measurements and connect the measurement results by a mathematical function, which can then be considered a general law, because physics is convinced that we can discover general laws through individual measurements. This procedure is called induction.

The method of physics to understand the world thus consists essentially in measurement and the search for the mathematical laws that fit these measurement results. To do this, phenomena that we observe must be transformed into numbers, which are then identified as representatives of this law. They must always be laws that precisely define the temporal and spatial change. This is established by definition because physics is convinced that the whole world is a machine and is therefore mathematically describable.

So we need a procedure to generate numbers from observations. That is measurement. And we need a mathematical law that fits these numbers.

## 1.1 Measurement

But what is a measurement and how do we carry out a measurement in practice?

First of all, it is clear that the result of a measurement must be a number if the measurement result is to be part of a mathematical law. But how do we get from the observation to a number?

Well, what numbers were already observable in the world before physics existed? First of all, in a world full of objects, we can count these objects. There are five glasses on the table, there are three trees in the garden, the house has seven windows. So counting objects is a measurement and the result of the measurement is a number.

We can also compare the distance of objects and use a scale to arrive at a number. This tree is three steps away from us, another tree is seven steps away from us. We can take the length of the step as a measure, or the length of the foot, or the distance from the elbow to the tip of the little finger. The measurement should ideally have the property that it does not change while we are measuring. The measurement then consists of counting how many times the measurement fits into the distance to be measured. To make the measurement even more precise, we can take a certain object as a measure of length, which has the property that its length is relatively independent of external influences. This gives us a measure of space and a first measuring device.

How do we now get from the observation that objects change their location in space at different rates to a measure of motion? We can see very different movements in the world. Trees moving back and forth in the wind. People and animals moving at different speeds. *Movement* means that objects change their position in space and that this change happens at different speeds. So we need a measure for this change – and that is physical time (see section 2.3). With our measure of time we can then measure movement, because the greater the distance an object has travelled in a fixed unit of time, the greater its speed.

We get a measure of time by comparing the movements of different objects, just as we did when measuring space. We take one movement as a reference and compare the other movements with this reference movement. Movements that differ from all other movements are suitable as references. These are the movements of the sun, the moon, and the stars. These movements give the impression that they run very evenly and that they are completely unimpressed by any processes on Earth. It is therefore natural to take the movement of the sun, moon, and stars as a measure of time. When the sun has travelled a certain distance, we say that a certain time has passed. The measure of time is therefore a fixed distance travelled by the sun. But in order to be able to assign a number to this measured time, we have to measure the length of the path again with our measuring device for length.

Experience shows that there are other movements besides the sun that are very regular, like those of pendulums and hourglasses. And we find that a certain distance travelled by the sun always corresponds to a certain number of oscillations of a pendulum or a certain level of an hourglass. Therefore, it is obvious to define the unit of time independently of the reference object.

Since the measured length is different for these different reference objects, we have to determine for each reference object which change in length corresponds to which time.

Instead of saying the sun has travelled such and such a distance or a certain pendulum has swung back and forth such and such a number of times or the level of the hourglass has increased by such and such a number of millimetres, we say: an hour has passed. And we give all reference objects the name *clock*.

A clock is therefore a system that has the property of moving as uniformly as possible or, more generally, of changing spatially as uniformly as possible. This spatial change is then transformed into a number with the help of a measuring device for space, and this number is transformed into a number according to a rule that applies to the respective clock, which is then a measure of time. A clock is therefore our measure and our measuring device for time.

We now have everything we need to measure the speed of an object. We mark the position of our clock and also mark the location of the object. Then we observe our clock, and when it has travelled a distance we have determined, we mark the new location of the object. Finally, we measure the distance the object has travelled between these two marks. We then have as the result of our measurement firstly the distance that our clock has travelled and secondly the distance that the object has travelled. Since the distance our clock has travelled is the measure of time and the longer the distance our object has travelled, the greater its speed, the speed of our object is the distance the object has travelled divided by the distance our clock has travelled.

We have achieved our goal and can now assign a number to each movement of an object by measuring this movement.

But it is worth taking a look at how exactly we have achieved this. We have found a way to measure space by creating a measure of length. And we have discovered a way to measure change by inventing a measuring device called a *clock* that lets us measure time. In both cases, however, the result of the measurement is a length. That is, if our measuring device is a length measure, then we get a length as the result of the measurement and we also call the result a length.

However, if our measuring device is a clock, then we also get a length as the result of the measurement, but we call the result time. The reason is that we have constructed the measuring device *clock* in such a way that the distance travelled by the clock is the measure of time. The actual measurement of the clock is therefore a length measurement. The distance travelled by the clock is measured and converted into a number according to a rule that is valid for the respective clock, which we call time.

So far, our measurements have always consisted of counting. We can count objects, we can measure lengths by counting how many times a measure of length fits into the length we want to measure, and we have a measuring device for time called a clock that presents the change in the world as a length that we count again with our method for measuring length.

Encouraged by its success in describing the movements of objects, physics attempts to quantify all perceptions by the method of measurement.

In the temperate zones of the Earth, we perceive that it usually gets warmer and warmer during the day and colder again at night. We call this sensation of cold and warmth "temperature" and again try to assign a number to the change in temperature.

To do this, we must again look for a reference object that changes as uniformly as possible when the temperature changes. Mercury, for example, is such an object. If we have mercury in a thin glass tube, it will expand when it gets warmer and contract when it gets colder. So the thread of mercury becomes longer or shorter. This change in length is then the measure of the change in temperature, and we get the number by measuring the length.

If our measuring device is a thermometer, then we get a length as a measurement result, measure this length and call the result of this length measurement *temperature*.

Here too, as with time measurement, we have to carry out two steps. First we create a length and then we make a number out of this length.

We can apply this method to other qualities that we perceive. And we will always look for reference objects that change spatially when this quality changes. We then measure this spatial change and get a length as a measurement result again, to which we assign a number according to the known method.

We measure weight, electrical charge, blood pressure, elasticity, defective vision – and the measurement result is always a length. Depending on which measuring device we used to measure this length, we call the result *length* or *time* or *temperature*, etc.

Since we always get a length as a measurement result, the measurement result is not self-explanatory. If we do not know what a measuring device was built for, then we cannot know what the measured length means. That is why measuring devices also have a scale that indicates what a measured length means.

There are also other methods to make quantifiable measurements. We can do tests, count the correct answers and calculate the IQ score from that, for example. Or we can do surveys and count the answers that fit into a scheme we have set up to find out the percentage of approval for a policy measure.

We always take delimited entities that we can count, or we create delimited entities for ourselves that we can count. And then we apply a calculation method that we have determined in order to get a number from it as a measurement result.

In our world of perceptions, there are very many qualities that we can measure. But there are also qualities that we cannot measure. Love, suffering, happiness, beauty, etc. are perceptions for which there is no measuring device. Why is that?

We have seen that a measurement always consists of finding a reference object. An object that changes spatially when the quality to be measured changes. But there are simply no objects that change spatially when we perceive changes in love, suffering, happiness, or beauty.

Because we always find a reference object for qualities that we can measure, we call these qualities objective. Qualities for which we cannot find a reference object, on the other hand, we call subjective. We then mean that objective qualities that we can measure are real, while subjective qualities that we cannot measure have no reality.

So the only way to quantify the world is to count it. Either we count objects or events directly or we build a measuring device that delivers a length as a measurement result, which we then convert into a number according to our length measurement method.

A measurement is therefore never a mere perception. A measuring device is conceived and then built by humans to fulfil a specific purpose. In order to understand the result of the measurement, we need to know for what purpose the measuring device was built. The measurement of the world is therefore not a determination of facts, but a man-made interaction with the world. The observer is not a neutral spectator but an actor who interacts with the world and interprets the result of this interaction as a measured value.

In the considerations so far, we have moved completely within the imaginary world of classical physics. We have used the terms *object, observer, measurement,* etc. as they are taught in every textbook of classical physics. That is, we are not talking about quantum mechanics or relativity here, and yet we have established that what physicists do and what they think they do are not the same thing.

Even the simplest measurements which fit easily into the worldview of classical physics are not mere perceptions of facts. All measurements, even in classical physics, cannot be understood at all without people involved.

## 1.2 Machine

The investigation into what a measurement is has yielded a rather surprising result. Measurements provide numbers, but the meaning of these numbers is anything but self-explanatory. There can be no question of facts of the matter. We have to interpret every measurement result – that is, we have to decide what the measurement result means. So the only fact about an experiment is that we always get the same length as a measurement result quite reliably. No matter how often we repeat the experiment and no matter who carries out this experiment, we always get approximately the same result.

This is an important insight, but initially it means nothing other than that we can construct an experimental setup that has a controlled and always the same effect on the world. Physics is thus in possession of technical knowledge to construct apparatuses that deliver pretty much exactly the same results.

But why does physics believe that this ability helps us to understand the world? Why should constructions that reliably do what they are supposed to do explain to us why the world is the way it is, or appears to us the way it appears?

The reason is that physics believes that the world is a machine.

The more machines we devise and the better we understand how complex machines work, the more we believe we will discover the secret of the *world* machine. By building more and more new machines, we learn more and more precisely what resistance the world offers; what conditions have to be fulfilled for these machines to function. The goal is to build all conceivable machines and thus find all the mechanisms that are possible in the world, in order to ultimately understand the *world* machine.

But what does building a machine have to do with describing a falling stone? Why should finding a mathematical law for falling stones help us explain the world? Just because we assume that the world itself is a machine?

In order to understand why physics believes it can gain knowledge about the world through this approach, we must first take a closer look at what a machine is.

According to Wikipedia, a machine is *a physical system using power to apply forces and control movement to perform an action.* A machine therefore consists of parts that move relative to each other according to certain rules. These rules describe the mechanism of the machine and explain why it does what it does. If we know all the mechanisms of the machine, then we have fully understood the machine.

Because machines have these properties, we can find mathematical laws to describe them. When we switch on a machine, it runs exactly according to plan, just like a mathematical law where everything is fixed.

We can understand the mechanism of a machine without knowing who created it and what purpose the creator had in mind when he created it.

So we examine the parts of the machine and find the mechanical regularity according to which the parts interlock and from this we can deduce the mechanism of the machine. If we know the purpose of the machine, we are not only able to investigate the *how* it works, but also the *why* for every component of a machine. The reason why a component of a machine has this or that function is because it ensures the functioning of the machine as a whole. If we know the purpose of a machine, we can understand why a particular component works the way it does. Then the *description* of the function becomes an *explanation* of the function.

So by interacting with a machine, we understand how it works, because a question to a machine always provides a mechanical answer. An interaction with an unknown object initially provides few clues about the nature of that object. However, as soon as we know that the unknown object is a machine, we can understand how the machine works from the reaction to our interaction.

Since a machine is no more than the sum of its parts, we can infer the machine as a whole from an understanding of the parts. Therefore, the law of falling stones reveals to us a part of the *world* machine.

If we now look at experiments and measuring devices, they have the character of machines. As we have seen, measurement results are not self-explanatory, but with the belief that the world is a machine, the measurement result can be explained as describing a mechanism. The assumption that the world is a machine and that we can interpret the laws we find as mechanisms is sufficient to have an explanation of the world. The meaning of a measurement result is then no longer freely selectable, because measurement results must always be part of a mechanical law.

# 2 The philosophical foundations of physics

Physics believes that its discipline is about facts and that this distinguishes it from all philosophical and religious belief systems. We can believe in the Bible's creation story or in Plato's doctrine of ideas, but we can also choose not to. Physics, on the other hand, provides us with facts. And with facts, we cannot choose to believe in them or not, because facts are self-explanatory. Physics translates its facts into mathematical laws that can rival the greatest thinkers of mankind in formal rigour.

But physics comes from a philosophical tradition. The first physicists were philosophers in their own right. Isaac Newton's groundbreaking work is called *The Mathematical Foundations of Natural Philosophy.* Therefore, physics has also adopted many ideas, assumptions, and concepts from philosophy that are by no means supported by facts.

Since physics relies on its facts and its mathematics, let us take a closer look at the use of some central physical terms to see whether physics applies the same rigour of thought to these terms as mathematics does.

## 2.1 Insights from machine theory

Galileo shifted the focus from *why* to *how.* Not *Why does the world work?*, but *How does the world work?* But *How does the world work?* initially only provides a description.

Physics believes that a description of the world becomes an explanation of the world if we assume that the world is a machine. Because physics believes that the world is a machine whose mechanism must be understood in order to comprehend it, physics is also mechanics – classical mechanics or quantum mechanics. All the mathematical apparatus of physics and all the interacting with the world by means of experiments and measurements serve the task of understanding the mechanism of the *world* machine. For as soon as we assume that the world is a machine and experiments and measurements serve to recognize the mechanism of the world, measurement results can no longer be interpreted arbitrarily, but must be data of a mechanism that we must find and which we can describe by mathematical laws.

And physics is very good at always building new machines and describing how they work. All so-called exact sciences are looking for mechanisms that can be described with mathematical laws. Because if the world as a whole is a machine, then all its parts are building blocks of this machine and must obey mechanical laws. Nature, man, society, the market. They are all part of the *world* machine and can therefore be described with mechanical laws that represent the temporal course of a variable.

So now we look at the consequences of the idea that the whole world is a machine.

A machine does what it does in a precisely calculated way. That is what makes a machine so valuable. It doesn't get tired, it doesn't change its mechanism of its own accord, it just works.

A real machine wears out over time and must be regularly maintained and repaired, otherwise it will eventually break down and no longer function. An ideal machine works for all time, as designed, and follows the mathematical laws that describe it. For Newton and Leibniz, an important question was whether the *world* machine works for all time as God created it, or whether God must occasionally intervene in the mechanism of the world to keep it running. According to Leibniz, God created a perfect machine; according to Newton, God must occasionally intervene.

The idea that the *world* machine needs to be maintained has now disappeared from physics. Once created, the world runs by itself for all eternity. So if the world is an ideal machine, the future and past are fixed and can be described by mathematical laws.

This is where the idea of a Big Bang in physical cosmology comes from. Since physics believes it can observe an expansion of the universe, it concludes that if the universe is getting bigger, it must have been smaller earlier. And because of the rate of expansion, it can then go back in time and calculate when the universe was only the size of a dot.

In doing so, however, it makes assumptions that are in no way derived from observation. For example, physics assumes that the laws of nature that apply today near the Earth are valid throughout the universe and were also valid at the time of the Big Bang. From the beginning of the expansion until today, nothing and no one has disturbed the universe in its expansion. It was therefore already certain at the beginning that it would look the way we perceive it today.

For physics, therefore, the only creative act is the creation of the universe itself. Everything that follows is predetermined. Past and future are fixed. Any further creativity or creation is excluded. And since we humans are part of this universe, we cannot have free will and cannot be creative. Creativity cannot exist because nothing new can be created if the course of the world has been fixed from the beginning. And since everything is predetermined, there can be no coincidences either. What we call coincidence is then only a lack of knowledge.

Machines have other very important properties. Machines are built by us to fulfil a purpose. There is no such thing as a machine without a purpose. There is always a reason why we build a machine. Our consciousness recognizes that we can better manage certain processes in nature with the help of a machine. Then we think about what the machine should look like in order to do this task. We build this machine and change it until it does what it is supposed to do. So a machine does not come into being by itself. Machines always have one or more creators, and machines always have a purpose.

When we build a machine, the creation of the idea for the machine is something quite different from the execution or implementation of the idea. The implementation of the idea is an act that must be carried out according to the specifications of natural conditions and can thus be learned as a technique as well as being carried out mechanically. The creation of the idea for the machine is a creative act that creates something new and is therefore unpredictable and consequently unlearnable; we can therefore also not mechanically perform and automate this act of creation.

So if the whole world is to be a machine, then there must also be a creator of this world. But if modern physics dispenses with the creator God, then the world cannot be a machine either. And if the world is not a machine, then mathematical laws that describe mechanisms cannot explain the world to us. The construction of ever new machines with the aim of understanding the world is then a futile effort.

Finally, a few thoughts on how we find out how a machine works. We can observe and examine a machine from the outside. It is not impressed in any way by the examination. We can take out all the parts, examine their function, and then put them back in without changing the machine. This is where the idea comes from that an observation of nature has no effect on nature. We can do experiments and get results and then pretend that the results would exist even if we had not done the experiments, because a machine does not change when we examine it as long as we do not damage or destroy it.

But we cannot observe and examine a machine if we ourselves are part of that machine. We have to look at the machine from the outside. If the whole world is a machine, then we are part of this machine and therefore we cannot observe and examine the *world* machine.

So if we want to hold on to the idea that the world is a machine, then we must also believe that there is a creator of the world. And we must believe that everything in the world is predetermined and that we humans have no free will and cannot be creative. But what have we gained from that?

We can describe machines with the help of their mechanism, but only when we know their purpose can we explain them, because the *why* of their existence does not follow from the *how* of their function.

So even if the world were a machine and we could actually figure out how it works, knowing how it works would not help us understand what purpose it serves. We cannot know what idea the creator of the *world* machine had when he created it, and we cannot know why he built this machine exactly the way he did. The description of a machine is never an explanation of a machine.

Descriptions are something very useful. In trades, in technology, and in engineering, there have always been descriptions of how best to carry out the respective projects. But if science wants to be more than a trade or an engineering art, then it must not only describe effects in nature, but also explain them. Then elementary particle physics must be more than an instruction manual for building linear accelerators, and astrophysics more than an instruction manual for building telescopes.

This also applies to all other sciences that believe they can explain anything through experiments and measurements. There cannot be a mechanical explanation for phenomena in the world. And there is also increasing doubt about the dogma of machine theory.

Researchers in the exact sciences are no different from professionals in engineering and the trades. They have a knowledge of *how* the world reacts when we act on it, but they do not know *why* it reacts that way. And mathematics, which is so wonderfully suited to describing machines, is no help to us in our search for explanations.

We cannot overestimate the value of the rational scientific approach when it comes to solving problems and finding models to describe them. But these models have no claim to truth, and they cannot and do not explain the world to us. But if physics only has models, then the cosmology of physics is ultimately no different from other creation myths. We can believe in it – or not.

A science that wants to explain the world to us must provide us with more than descriptions and models.

## 2.2 Objects in space and time

We live in a world that is constantly changing and in which the most diverse perceptions affect us. We perceive heat and cold, we feel pain, we are hungry or thirsty, we feel resistance, we are afraid or feel joy. All these perceptions have the property that we can evaluate them: we call them qualities. In the following, *quality* does not mean a value-neutral property, but an assessable property (goodness).

Besides these qualities that we perceive, there are also things or objects like stones or tables or chairs that we perceive in a different way. We see them, hear them, or feel them, but they are different from the qualities. They appear solid to us, they can move, they are of lasting duration. But above all they are value-free, because we don't say this is a good stone or a bad stone. We call these things which we cannot evaluate "objects" and distinguish them from qualities. Of the qualities we perceive, there are some that we associate with an object. We can evaluate the smoothness of a stone, the grain of a wood, etc. and thus assign a quality to an object. But we also perceive qualities like love, joy, or pain that we cannot easily assign to an object.

When we are born, we cannot recognize objects. We can only perceive certain qualities, like hunger or cold, but we do not recognize tables and chairs. In the course of our development, we learn to perceive objects; that is, we learn to order the variety of appearances.

After we have learned to perceive objects, we can see these objects behind each other, next to each other, or on top of each other. We see that these objects have an edge – that is, they are limited – and we see that there are empty spaces where there are no objects. This is the first time we create the idea of space.

But the change in our world is still that these objects appear and disappear.

The next step in our development is to give these objects an identity and that we can remember them. The ball that disappears under a table and then reappears is not a new object, it is the same ball. And it hadn't disappeared at all, it was just hidden by the table. This means that the ball must obviously have moved if it disappeared under the table and then reappeared.

So as soon as we have a memory and give objects an identity, then objects no longer simply appear in different places in space, but they move from one place to another and they were also in all the places in between.

After we have assigned a movement to objects for the first time, we can see that objects move very differently. They can move fast or slow, change their speed, and move in such a way that they always come back. With some objects we can easily influence the movement, while other objects move independently of us.

The sun, the moon, and the stars have a special status, as they move very evenly and their movements seem to be completely unimpressed by the movements of other objects.

So we have the possibility to compare the movement of objects in our environment with the movement of these particular objects.

But we do not only have the possibility to compare the movement of objects that surround us with the movement of the sun. We also have the possibility to compare all changes in the world we live in with the movement of these particular objects. We can compare the changes in vegetation with the changes in the stars or the change in the feeling of hunger with the movement of the sun. This gives us an organizing principle for the changes in the world and thus *time*.

The idea of *time* is therefore based on the comparison of the changes we perceive with the movement of objects whose movements show greater regularity compared to most other movements – for example, the movements of the sun, moon, or stars. Their movements appear to us to be unaffected by the other changes in the world.

On this basis, ideas such as development are then possible; that is, changes can be evaluated. Concepts such as *beginning* and *end* or *birth* and *death* are then possible. For the idea of time is linked to the idea of movement. Mere changes in phenomena without the idea of movement do not lead to an idea of time.

As long as there are no objects that move, there is no time, no philosophy, no science. The time of science is linked to the movement of objects. Everything we call objective is based on objects. Objective time is based on the movement of objects. Therefore, in the following section we will take a closer look at objects and their movement.

### 2.2.1 Objects

When we observe the world, we assume that what we observe is not influenced by our observation. The moon is there even when no one is looking.

If we look at an object, it has a certain appearance. We can walk around this object, then it might look different. From one side it might be green, from the other side blue. We can walk around the object multiple times, sometimes to the left, sometimes to the right, and each time the object will have the same appearance when seen from a certain place. So after we have walked around the object a few times, we know its appearance from all positions and already know in advance what it will look like when we look at it from a certain position. It is therefore natural to assume that it will look the same in all these places when we are not looking at it.

This assumption is generalized in physics to the effect that the properties of all objects we can observe would also have been there if we had not observed them. The property that is observed then does not depend on the observation, but is assigned to the object as part of its essence. The observed property thus belongs to the object and is merely taken note of through the observation.

But what exactly do we mean when we say that the properties of the object are there even if we do not perceive them?

Let's take a green, soft cloth made of velvet as an example. It is green for us because we see it that way. It is soft because we can touch it and feel its softness with our sense of touch. *Green* and *soft* are qualities we recognize because we know them. What would be the point of saying that the cloth is green and soft if there were no people who could see and touch it? Now we could say: "Okay, the cloth may only be green or soft for us humans, but there is no doubt that the object we call cloth exists as such."

If we put the cloth in a cupboard and take it out again later, it is still the same cloth, and we also assume that it was the same cloth while it was lying unobserved in the cupboard.

We might now be willing to admit that the properties of the cloth such as *green* and *soft* do indeed have something to do with our perception. But we might now argue that while *green* and *soft* are properties that depend on our perception, there must be something there that causes this perception of *green* and *soft*. Something about this thing makes it look green and feel soft when we perceive it. But what could that something be? It would have to be something that we don't recognize, but that exists nonetheless. It would be what Kant called the *thing-in-itself*.

But how do we imagine that something exists but cannot be recognized by us? If we say that there is something that really exists but cannot be recognized by us, then that would be a recognition for us. We could then say that we have recognized that there is something that is beyond our recognition. At the same time, however, we also say that it is not recognizable. But with that we have a logical contradiction, because if it is not recognizable, then we cannot recognize that it lies outside our recognition.

So we cannot leave the realm of recognition, because even if we say that there is such a thing as the *thing-in-itself,* but that we cannot further recognize it, then that would also be a recognition. Because even a recognition that we *cannot* recognize something is a recognition.

Consequently, we cannot avoid the recognition that whenever we say that something *is,* that then also means that it is recognized.

So if we subtract perception from objects, then nothing remains.

In Western philosophy, the first to recognize this was the Irish philosopher George Berkeley: *esse est percipi* (to be is to be perceived). However, Karl-Heinz Brodbeck points out in his book *Der Zirkel des Wissens* (The circle of knowledge) that the Cittamātra School of Buddhism had already described this insight in detail one and a half millennia earlier.

But if only what we perceive of objects exists, then a perceiving subject always belongs to a perceived object. Consequently, an objective view is only possible for subjects. An object without a subject does not exist. Objects without subjects have no intrinsic being. But the reverse is also true. Perceiving subjects without perceived objects also have no intrinsic being.

We have to imagine the logic of the relationship between perceived object and perceiving subject as with the conceptual pair of *deed and perpetrator.* Just as there is neither a deed without a perpetrator nor a perpetrator without a deed, there is no perceiver without a perceived and no perceived without a perceiver. Both always arise simultaneously and do not exist without the other.

With this we also understand why an objective description of elementary particles cannot work. Since we cannot perceive elementary particles, there is no perceiving subject for elementary particles. But if there is no perceiving subject for elementary particles, then elementary particles cannot be objects.

### 2.2.2 Movement

The question of the location and speed of objects is also interesting.

There is an apple on the table in front of me. The question of where the apple is and at what speed it is moving is easy to answer. It is 10 cm from the left edge of the table and 5 cm from the front edge of the table and it is not moving – that is, its speed is zero.

In fact, however, I am sitting in a train that is moving at about 150 km/h. For an observer standing on the railway embankment and watching the train pass, the apple is moving just as fast as the train, i.e. at 150 km/h. As a result, the location of the apple also changes constantly, as it moves away from the observer on the railway embankment together with the train. Obviously, the assessment of where my apple is and how fast it is moving depends on where my location is from which I am making my observation. So the question of where the apple is and at what speed it is moving is a question of point of view. The term used in physics for point of view is *frame of reference*.

But at what speed is the apple actually moving at? And is it moving at all?

In the objective world of classical physics, objects have the property that they move at different speeds in different frames of reference, but after returning to a previously adopted frame of reference, the objects are just as fast there as they were before. We can infer from one frame of reference to other frames of reference and create mathematically exact transformation functions that transform the speed of the objects into other frames of reference, which we can then also observe. No matter which frame of reference we choose, our objects always have the velocities we have calculated.

Thus, we can also assume that the location and velocity of an object are properties of the object without contradicting observation.

This is the starting point for Newton's theory of mechanics. Objects move in themselves and their true motion is the motion in absolute space. Even though the observed motion of an object depends on the observer's frame of reference, this observed motion is mathematically a transformation of the true motion that the object has in absolute space into the observer's frame of reference. So if our frame of reference is in absolute space, then we are observing the true motion of objects. If we move relative to absolute space, then the transformation law, called the Galilean transformation, tells us how the motions of objects change. It does not matter if we know which is the frame of reference in absolute space. Whenever we have two frames of reference moving relative to each other, the Galilean transformation tells us how the velocities of the objects change when we move from one frame of reference to another.

However, all attempts in physics to find out through experiments and measurements what the true, absolute speed of an object is have failed. After Einstein published his special theory of relativity in 1905, the idea that there is an absolute space in which all objects have their true velocity was finally dropped. One point of view is as good as any other. The question of the true point of view from which we can measure the true motion of an object is meaningless. This realization is called the principle of relativity.

But if there is no true motion of an object, then velocity cannot be a property of an object. Then speed is only a relation. We can say that an object is faster than we are, or slower, or the same speed, but the question of how fast it really is is meaningless.

If movement is relative, then movement is not a property of an object, but only a relation. Size, for example, is a relation. If we have two tomatoes and one tomato is smaller than the other tomato, then this being smaller is not a property of the smaller tomato, because we always need a second tomato to be able to say that one tomato is smaller than the other. If we eat the bigger tomato and only have one tomato left, then it makes no sense to say that this tomato is smaller. Smaller than what? And it also makes no sense to ask how big this tomato is in itself.

If movement is not a property of objects, then there is no absolute time. Time is the comparison of objects that move differently (see section 1.1 *Measurement*). And if the movement of objects is dependent on location, then time is also dependent on location.

But physics still believes that motion is a property of objects. It knows that we cannot know how fast an object is moving, but we always have a point of view from which we can measure the speed of an object, and we have our transformation law that tells us exactly how great the speed of our object is from a point of view that is moving relative to it. So even though location and velocity are not properties of objects, physics still assumes that we can describe all motion of objects by what is called the equation of motion. In this equation of motion, every object *has* a location in space and a certain velocity at every point in time. Although relativity theory has shown that it is pointless to assign location and velocity to an object as a property, physics does it anyway, and accordingly it deals with some paradoxes.

What does it mean when objects *have* a location? And if they *have* a speed at that location? If objects move from A to B, *are* they also at all locations between A and B?

Intuitively, we would say that an object moving from A to B follows a trajectory and, of course, has been to all the locations between A and B.

To answer the question of whether the object was actually also in all the places between A and B, we can consider how long it stays in each of these places, because if an object is to be in a place, then it must be there for at least a short moment.

Let's imagine that we hit a billiard ball on a billiard table so that it rolls across the table at the speed of one metre per second. If we have two marks on the table that are one metre apart, then the ball takes one second to roll from the first mark

A to the second mark B. Or in other words, the ball is between A and B for one second.

If we consider a smaller interval between A and B that is ten centimetres long, i.e. one tenth of the length between A and B, then the ball is also only in this smaller interval for one tenth of the time. The smaller the interval we consider, the shorter the period of time during which the ball is in this interval.

And now comes the surprise. If we want to know how long the object stays at a certain point on the curve, then we have to make the interval smaller and smaller until it has a length of zero. Then we no longer have an interval, but a point. But if the interval has the length zero, then the time the object spends in this interval is also zero. And an object that stays in one place for zero seconds cannot have been there.

So if an object moves, then it cannot be *in* one place. And vice versa it follows: if an object *is* in a place, then it cannot be moving, but must be at rest.

This insight is by no means new, but is one of the oldest insights of humankind. Zeno of Elea already made these considerations more than 2,000 years ago.

But how do we deal with such a recognition? We now seem to have the dilemma of either believing that objects are in places and cannot move or move but cannot be in a place.

What are the solutions to the dilemma?

Zeno held that the object was *in* the places between A and B and therefore could not move, that movement was therefore an illusion.

Physics believes it has found the solution to this problem in the mathematical limit concept and shows that in a space–

time continuum one can find a formal calculus that still assigns a velocity to a moving object even in an interval of zero length. Our conclusion that an object that *is* in a place must be at rest would be disproved. With this solution, the dilemma would disappear.

Apart from the fact that mathematical calculations are abstractions that have only a limited relationship with reality, physics arrives at a solution that is not even conceivable. What sense does an event that lasts zero seconds make? I can truthfully say that I was the Pope for zero seconds. But if I claim that I don't mean that I was never the Pope, but claim that I really was the Pope, even if only for zero seconds, then it makes no sense.

The problem is that physics believes that there are points in space and points in time. But a point in time has a duration of zero seconds. A duration, no matter how short, cannot be zero.

And you can't measure a point in time either. In order to measure a time, our clock must have moved, because time measurement is always a length that our clock has travelled. And you can't measure a speed at a point in space either. To be able to measure speed, the object must have moved a little. Since the reality of physics is always a measured reality, the speed of an object at a point in space cannot be a physical reality.

That is, after all, precisely what the abandonment of the notion of absolute space means. So the solution to the dilemma is to recognize that both place and velocity cannot be properties of objects and therefore there is no dilemma. Objects *have* no place and no speed.

So although physics has come to the realization that there is no absolute space, it has obviously not understood the significance of this realization. If physics had understood its own insight, it would know that we cannot explain reality with equations of motion.

There are other examples that show that physics has obviously not understood its own concept of motion.

This includes the statement of physics that the universe is expanding. But if the universe is the totality of everything that is supposed to exist, then it cannot have an edge. Because if it had an edge, then there would still be something behind this edge and then the universe would not be the entire world. But if the universe has no edge and is infinitely large, then it cannot expand. Since it is already infinitely large, it can hardly become even larger. And if the universe is infinitely large, then it can't have been smaller in the past.

There is another difficulty with the idea that the universe is expanding. If motion is a relation between object and frame of reference, then what is the frame of reference for the expanding motion of the universe? At best, we could say that observed from Earth, it looks as if the universe is expanding. We cannot know whether the universe is really expanding, because it is a meaningless question.

We can also believe that we are just an insignificant existence on the edge of the Milky Way, but it is equally possible for us to believe that we are the centre of the universe. Because points of view are relative according to the theory of relativity. We can also believe that the Earth is the centre of our solar system. The same physics would apply. Only the equations of motion would be much more complicated. But who says that equations of motion have to be simple?

Physics places so much emphasis on being an empirical science, and yet at the same time believes in unthinkable abstractions such as space and time points, which are, after all, beyond any empiricism.

Cosmology is not an empirical science either. Otherwise, one would have to examine multiple universes in the laboratory. Putting together recognized building blocks is not empiricism, but thought experiments. Exactly what physics accuses philosophy of. You can deduce all kinds of things from background radiation, but it's not empirical.

Cosmology also has to face the question of what time should have been in effect at the Big Bang. If physical time is based on the comparison of moving objects and if the movement of objects depends on the observer's point of view, how can there have been a time before there were observers with points of view?

In quantum physics, too, elementary particles are attributed movements that make no sense. For example, some elementary particles are said to be point-like, i.e. they have no spatial extension, and yet they are said to have an angular momentum, i.e. that they rotate around themselves. But what is supposed to rotate around what? A point without spatial extension cannot rotate around itself. But if one nevertheless insists that a point-shaped particle has angular momentum, this leads to the alleged action-at-a-distance conundrum of entangled systems, also called the Einstein–Podolsky–Rosen (EPR) paradox.

### 2.2.3 Summary

Physics as the attempt to explain reality through the spatio-temporal motion of objects that exist independently of observation has failed, because objects without an observer or subject are inconceivable and motion as a property of objects contradicts the principle of relativity.

## 2.3 The observer

With the theory of relativity and quantum mechanics, the concept of the observer is explicitly introduced into physics.

But what is an observer? How do we recognize an observer? If the observer is a human being, what distinguishes him from a human being who is not an observer?

Let's imagine we are sitting on a bench in the park and see a woman watching a squirrel hop across a meadow, climb a tree, and disappear into the branches. Then we would probably describe this scene as having observed a woman observing a squirrel. So we would have observed a scene and observed an observer.

However, this description, which is so commonplace for us, is paradoxical on closer inspection. When we speak of an observer, we make a distinction and say that there is an observer and there is something observed. The observer and the observed are not the same, they are different. But this means that we cannot observe an observer.

If we observe the woman who is looking at the squirrel, then this woman cannot be an observer. Since we observe her, she is the observed, because everything we observe is always the observed.

If we could observe an observer, then there would be no difference between the observer and the observed. Then the human being whom we observe would be both something observed (we observe her watching the squirrel) and an observer, and the distinction between observer and observation would be removed.

But we cannot first presuppose a difference between observer and observed and then abolish it. That would be paradoxical.

Therefore, the question of whether an observer influences the observed or not cannot be answered, since the observer is not observable. An answer to this question presupposes that an interaction between the observer and the observed could be observed and that either an influence could be detected or not. However, this would mean that the observer is observable in his interaction with the observed and thus no longer an observer, but something observed.

According to relativity and quantum mechanics, the world of physics is an observed world. But if we introduce the concept of the observer into physics, then the world of physics is no longer the whole world, but only an observed world in which the observer is absent and physical statements about an unobserved world make no sense. The universe of physics is then only the observed starry sky and not the whole world.

Nevertheless, physics still believes that physical statements about an unobserved world are meaningful. But if the universe is said to have emerged from the Big Bang, who or what is the observer of this process? Who or what observed the formation of our solar system and the emergence of life on Earth?

Everything we know about the world from science are observations of science, and always science claims that everything it observes actually exists, and would exist even if it had not observed it.

Not only in physics, but also in other sciences, scientists often want to give the impression that their presence during the experiment had no influence on the result.

Psychologists create artificial situations, give their test persons some kind of task and then evaluate the solutions of their test persons in a strictly scientific way. They then claim to have gained valid statements about human behaviour, completely ignoring the fact that the subjects might have behaved differently if they had not been observed.

Pollsters ask people on the street questions about their opinion on certain topics and then pretend that these people would have had this opinion even if they had not been asked. As if the way the question was asked had no influence on the answer.

Experimental physicists often go to enormous lengths to create an artificial situation for their experiments. The outside world must be completely isolated so as not to have any influence on the experiment. Then they find a reaction of nature and claim that they have found a property of nature that has nothing to do with their experiments. But every physical experiment is a complex of social action and corresponds to the construction of a machine. Every experiment has an idea as its beginning and a purpose as its end. In between, matter is used to create a mechanical construction that fulfils the desired purpose. Just like building a machine. Natural material is reshaped and worked on until it fulfils the desired purpose.

After the experiment, this whole social process is forgotten and reduced to the measurement result, and scientists then claim that they found something, when in fact they produced something.

## 2.4 Cause and effect

When we eat, our feeling of hunger disappears. If we take a painkiller, the pain disappears. When we drop a porcelain vase on a stone floor, it breaks.

In the course of our lives, we experience that many things and events are interrelated. When it rains, we get wet. If we get too close to a fire, we get burnt.

And when we have had these experiences often enough, we believe that there is such a thing as a fixed connection between things and events. We call these connections cause and effect and think that what takes place first is the cause and that what happens afterwards is the effect. First it starts to rain and then we get wet. The rain is the cause and our wet clothes are the effect. The cause is the reason for the effect, or in other words, the cause causes the effect.

We structure our whole life according to these contexts. If our alarm clock fails, we miss the bus and are late for work. Because we know this and assume that this is always the case, we may set a second alarm clock if we absolutely must not miss the bus.

So we gain the experience that there are certain connections that belong firmly to each other, and when we have had this experience often enough, we are convinced that this has not only been the case so far, but will always be the case in the future.

So we believe that there is a natural law that establishes these connections. And that is also the basic idea of science. Science assumes that there are regularities in nature that run according to fixed rules, and it makes it its business to describe these regularities with the help of mathematical laws. And if science finds mathematical laws that can describe these relationships, then it is also convinced that these laws have a universally valid character; that these laws can not only describe past events, but also apply to the future.

There are thus three ideas about cause and effect that form the foundation of the sciences, especially physics. Firstly, the cause comes before the effect and causes it; secondly, we may conclude general laws from individual observations; and thirdly, these general laws should apply for all time, i.e. also in the future. The idea that the cause causes the effect is called causality. The idea that we may conclude from individual observations to general laws that apply for all time is called induction.

In Western philosophy, the Scottish philosopher David Hume was the first to question these ideas that we take so much for granted.

If we believe that a certain cause always produces a certain effect, then we would have to somehow discover the effect in the cause. That is, if we examine the cause independently of the effect, then we would have to find a clue in the cause as to what effect it produces. But we cannot do that. There is no object and no something for which, without knowledge of its effect, we could predict what effect it causes. There is no logical reason for the cause to cause the effect.

When we see an open fire for the first time, we don't realize that we will get burnt if we get too close. We can't tell by looking at the fire that it has this effect on our skin. Only after we have had the experience a few times do we realize the connection.

So we have the experience that there is something we call cause and something we call effect, but without this experience we cannot discover the effect in the cause.

Now, one could argue that while we cannot discover the effect in the cause, that does not mean that the effect is not in the cause. Perhaps we have not searched thoroughly enough. Or we have searched with the wrong method. And perhaps that is why science has simply ignored these considerations of Hume.

But there is an Indian philosopher called Nagarjuna who lived in the 2nd century and is quite unknown in the Western world. He proved that the effect *cannot* lie in the cause at all. That the assumption that the cause produces the effect leads to logical contradictions.

The questions we must first ask ourselves are: "What *is* a cause and what *is* an effect?"

The crucial word is *is*. When we say there is a cause that produces an effect, we mean that a something exists that we call cause and that a something exists that we call effect. Cause and effect are entities that exist for themselves and are distinct.

This is where Nagarjuna comes in. He said:

> If things did not exist without essence, the phrase, "When this exists so this will be," would not be acceptable.

To understand this difficult sentence, we need to backtrack a little.

The first part of the sentence reads:

> If things did not exist without essence, ...

If we say that a thing really exists and that it exists independently of perception, then it must have an essence. *Essence* means that it exists out of itself (in itself). That it does not need another thing to exist. So the sentence means:

> Because things only really exist when they exist out of themselves and do not need another thing...

The second part reads:

> ...The phrase, "When this exists so this will be," would not be acceptable.

This means:

> ...If a thing exists in itself and consequently has an intrinsic being, then it cannot transform itself into another thing.

We can therefore translate Nagarjuna's statement into the somewhat more understandable statement

> Because things only really exist (in themselves) when they exist out of themselves and do not need another thing, a thing cannot transform itself into another thing.

Applied to cause and effect, Nagarjuna says:

> If cause and effect are something that exist out of themselves without depending on other things, that is, have an existence out of themselves, then the cause cannot become an effect.

At first glance, this is not obvious. Why can't the cause become the effect if cause and effect exist independently of each other?

Our everyday experience tells us something quite different. If we want to drive a nail into the wall, miss, and hit our thumb, it is quite painful. And of course we are convinced that the blow on the thumb is the cause of the subsequent pain in the thumb. First comes the cause (hitting the thumb), then comes the effect (pain in the thumb).

But if we take a closer look at this process, it is not so easy to see what exactly is the cause and what is the effect. When the hammer moves towards the thumb, we could say that the cause of the pain is the speed of the hammer before it hits the thumb. We could also say that the moment the hammer touches the thumb is the cause; or that the deformation of the thumb by the hammer is the cause; or that the release of nerve stimuli by the deformed thumb is the cause.

In fact, there is a continuous process here, and any stopping at a particular point with the assumption that the world at that point *is* in a fixed state that we can call cause is completely arbitrary and leads to the logical contradictions that Nagarjuna points out.

In our everyday experience, however, this is never a problem because we do not even consciously think that there is an entity called *cause* that has its own being, but simply perceive that there is a process where the hammer hits the thumb and shortly afterwards the pain occurs.

In physics, however, the situation is quite different. Physics, after all, follows the idea that there is a reality beyond our perception that consists of objects that exist in themselves, are in a fixed state at any given time, and influence each other through interaction. In this reality, causes are said to exist in themselves and independently of effect. But this is precisely what Nagarjuna has shown is inconceivable because it leads to logical contradictions.

For if something exists in itself, then its existence is not bound to conditions. So if cause and effect exist in themselves, then the cause cannot be a condition for the existence of the effect.

If we start from the idea that the cause exists temporally before the effect, then there is something that we call the cause and that really exists, while the effect is not yet there, so it does not yet exist, since the effect is supposed to come into being later. Then the cause is supposed to somehow produce the effect and in the process disappear itself. So first a cause exists and the effect does not exist. Then the cause merges into the effect. And finally the effect exists and the cause no longer exists.

How can we now imagine the transition of cause into effect?

If the cause disappears before the effect comes into being, then there would be nothing to cause the effect. So cause and effect must both exist for a certain time. During this time, the cause disappears and the effect comes into being. While the cause transforms into the effect, the cause must at some point cease to be different from the effect. But this contradicts the presupposition that the cause is an entity that exists for itself and must therefore be distinct from the entity of the effect. We cannot first presuppose that cause and effect differ as entities and then conclude that cause and effect no longer differ as entities.

We could now claim that there is a development according to which cause and effect first differ and then, in the course of time, no longer differ. However, this contradicts the assumption that cause and effect exist by themselves, which means that their existence cannot be linked to any conditions. A temporal development of the existence of cause and effect would, however, be a condition of their existence. Either cause and effect exist without their existence being linked to any condition - or not. Both together do not work.

So in order to continue to rightly speak of cause and effect, we have to get away from the idea that cause and effect are each separate entities. They only ever occur as a pair and are mutually dependent. Just as we have seen with the deed and the perpetrator. We cannot speak of the being of a perpetrator if there is no deed. The deed makes the perpetrator and vice versa.

It is not a matter of claiming that we are having a false experience when we say that because we hit our thumb with a hammer we feel pain. There can be no doubt at all about this connection. The problem only arises when we want to transform this if-then description into a cause-and-effect explanation.

The point is that physics wants to explain the world to us by claiming that there are such things as real existing states of the world which we call causes and real existing states which we call effects, and that the cause causes the effect. But this idea is logically unsustainable.

Another problem materializes for physics with the idea that the cause should cause the effect. Because if that is to be the case, then the two must be connected somehow. But we do not see this connection. Therefore, physics has to invent the concept of force. This force is supposed to be responsible for creating the invisible connection between cause and effect.

When we drop a stone, physics says that *gravity* pulls it down. The link between the cause that we let go of the stone and the effect that the stone lies on the ground afterwards is gravity. And gravity is also supposed to drive the planets in our solar system to move around the sun. But we don't see this gravity. We only see the motion. So Newton did not *find* gravity, but *invented* it in order to be able to maintain the illogical concept of cause and effect.

As if that were not strange enough, physics partly contradicts its own idea of cause and effect.

The characteristic of the exact sciences is to describe this cause–effect relationship mathematically and exactly. In the mathematically formulated world of physics, however, there are no causes and effects at all, since the equations of motion determine the past and the future. If the whole world were a machine created by God, then a state of the world at a certain point in time could not be the cause of a later state of the world. Since every state of the world was determined by the beginning, it is meaningless to claim that a later state was determined by an earlier state.

So, on the one hand, processes in the world are supposed to be explained by causes and effects connected with certain forces, and on the other hand, the world as a whole is supposed to have come into being through the Big Bang and to have developed exactly according to mathematical laws.

There are also situations where we talk about cause and effect, but it is actually about something completely different.

A good example of the improper use of the concepts of *cause* and *effect* is the image taken from chaos theory of the butterfly that is able to trigger a tornado by flapping its wings. Behind this is the idea that chaos theory teaches us that small causes can have a big effect.

Let's imagine that we had a mathematical model of the entire world, into which all data relevant to the weather were fed, i.e. the position and speed of all atoms and molecules of *everything* that covers the Earth's surface, inhabits it, or otherwise belongs to or influences the Earth's atmosphere. Then we could use this model to accurately predict the weather for all regions of the Earth for all time. Suppose we had two research institutes collecting this data, and one institute made a tiny mistake and forgot a single butterfly out of all the living things on Earth. Now, forgetting a single butterfly is such a tiny mistake that one would think that the difference would have no effect at all on the weather forecast. Now the achievement of chaos theory is precisely that it has shown that even such tiny errors have an enormous impact on the prediction of the weather. After a few hours, the predictions in the model with the butterfly and the predictions in the model without the butterfly would differ so much that a hurricane could occur in one model, while calm weather would be predicted in the other.

So the image of the butterfly is not about cause and effect at all, but about the temporal progression of very small differences. The butterfly is not the cause of the tornado. It is merely an example of two systems that differ very little. And the hurricane is also not the effect of the beating of the butterfly's wings, but an example of two systems that differ very much.

Something else does not fit with the picture that the butterfly is supposed to be the small cause and the tornado the big effect. There is no measure and no measuring device for the size of a cause or an effect. We have no idea whatsoever of the difference between small and large causes and the difference between small and large effects.

The fact that we nevertheless have the feeling in the above example of the butterfly and the tornado that it is a small cause with a large effect is due to the fact that we evaluate events. *Big* and *small* in this example are value judgements and not physical states. The event of *the butterfly flapping its wings* is an insignificant event and has a small impact on our lives, whereas a tornado is a very significant event that has a large impact on our lives.

## 2.5 Induction

Closely related to cause and effect is induction. Induction means that we conclude from a few, always identical correlations that we have recognized in the past that these correlations signify a general regularity and that these regularities will also exist in the future. Again, Hume realized that there is no logical reason for this conclusion.

Hume's argument can be split into two parts.

First, Hume deals with the question of how the connection between individual observations and a general law arises.

Whenever we drop an object, it falls to the ground. We then say that gravity pulls the object downwards. To be more precise, the Earth and our object attract each other because the law states that masses attract each other. However, since the mass of the Earth is much greater than the mass of our object, our object moves strongly towards the Earth and the Earth moves only a very little towards our object.

So we always have the same experience that objects we drop fall to the ground. However, we do not perceive gravity and the law that masses attract each other. The law is therefore something we invent, and it does not follow logically from our experiences. If the law were to follow logically from our experiences, then there would be a logical contradiction when we claim that there could be experiences that do not obey the law. But this is not so. Just because we have always had the same experience so far, it does not follow that someone else may not have had a different experience. It is possible that there are distant galaxies where this law does not apply. We cannot logically rule that out.

The second part of Hume's argument addresses the question of whether we may conclude from a law that has always applied in the past that it will also apply in the future. Again, such a conclusion is not necessarily logical. It does not contradict our experience that what we have always observed in the past and can describe with a law may no longer apply in the future. There is no logical reason why we could not have a different experience in the future.

Hume has shown us that the usual scientific procedure of inferring general laws which are also to apply in the future from individual observations cannot be logically justified. Natural laws are free inventions of human beings, and the assumption that natural laws are universally valid is merely a belief and nothing else.

## 2.6 Facts and opinions

Findings of science are considered facts or facts that distinguish them from mere opinions.

When we discuss controversial issues, occasionally the argument comes that something has been scientifically proven. Then this is considered a scientific fact and is different from arguments that are merely opinions, because a fact always has a higher value than an opinion.

Why is that? Why do we believe that facts are worth more than opinions?

Facts are considered self-explanatory – that is, they speak for themselves and they are not usually doubted. No matter what we believe and what ideas we have about the world, facts are true. Opinions, on the other hand, are something we believe in or are convinced of. We do not know whether an opinion is true. Opinions can gain in importance if they are shared by many, but it is always inherent in them that they can be changed at any time. A fact, on the other hand, does not change. And above all, there is no need to explain facts, because they speak for themselves, and they always remain facts, regardless of whether many or few people believe in them.

One then naturally wonders how it can come about that a scientific fact can be recognized as false when its truth is completely uninfluenced by human beings. As is well known, scientific progress consists of replacing earlier findings with new scientific findings and in recognizing former facts as false in the light of new scientific facts. How can this happen? How can a fact become false?

A fundamental error in the question of facts and opinions in science is in the judgement of what scientists do. Scientific experiments produce results that can usually be reproduced. Every time we repeat the experiment, it ideally leads to the same result. That is definitely a fact. But the result always requires interpretation. It never speaks for itself. Scientific findings always go back to some measurements that do not speak for themselves but have to be interpreted and therefore reflect the opinion of the scientists involved (see section 1.1 *Measurement*).

The statement that an electron has the charge $-1.602176634 \cdot 10^{-19}$ C is not a scientific fact. The fact is that one can construct an experimental setup that allows the interpretation that there are electrons that have this charge. But this interpretation is not logically compelling, because facts have the property that we cannot draw compellingly logical conclusions from them. They speak for themselves and only for themselves.

Scientific findings are therefore not facts, but opinions. Every scientific fact is the opinion of a sufficiently large group of scientists who have agreed on this common interpretation of a fact. And since opinions can change - even the opinions of scientists - it happens that scientific facts are replaced by new scientific facts.

In principle, there is nothing wrong with scientific findings being opinions, as long as this is also expressed and as long as the groups are not too small and the exchange in science as a whole functions. That means that the respective science as a whole represents this opinion. Moreover, all minority opinions should always be made public and be part of the published findings.

Even if the opinion of experts is to be valued more highly than the opinion of laypersons, it should still always be clear that these are opinions.

## 2.7 Is the truth in the action or in the things?

When we do science, we want to know the truth about the world we live in. Sciences are usually concerned with the objects of research that they study. Be it *the* elementary particles and *the* universe in physics, *the* societies in sociology, *the* brain cells and *the* brain in neuroscience, *the* cells in biology, and so on. And the aim of the sciences is to find true properties of their objects of research. In the natural sciences, the so-called exact sciences, objects of research are real things about which science wants to make true statements. It is always about properties that these objects of research *have*. An electron *is* a particle that *has* an electric charge. A brain *is* a complex organ that *has* a lot of neuronal cells. It is always about things that *are there* and *have* properties. So science seeks the truth in things.

It's just the way we approach things in everyday life. When we lift an object and notice that it is heavy, we also say that the object has the property of being heavy. We don't think that the object is heavy because we lift it up. The object *is heavy,* and it is heavy even if we do not lift it.

So when we look for truth in things, we assume that our observations have no influence on things. Our observations are then only a recognition of facts that are simply there, even if we had not observed them, because our observations neither change nor create things.

And in the early days of science, the objects of research actually gave us the impression that they were not impressed by our observation. Stars and planets, trees, mountains, houses... all the objects of our living world give us the feeling that they are not impressed by our perception.

Progressive technical development initially led to an expansion of science into areas that lay outside the immediate living environment, but which could be transformed into our sensory world with the help of technical aids such as telescopes and microscopes.

The problems with the idea that objects *are* simply *there* and *have* properties, i.e. the idea that truth lies in things, only came with the expansion of science into areas outside the realms of the living world. Be it the study of completely alien cultures or of objects that either move very fast or are very small.

In the 19th century, it became clear that the observer plays a greater role than previously assumed. The idea that an observer stands outside of what he observes became increasingly untenable. In the social sciences, the first thing to grow is the realization that a social science experiment is not without influence on the object of research. When we visit indigenous peoples who have never before seen people who look completely different, we cannot observe their behaviour as a neutral observer from the outside, because their behaviour is naturally changed by our presence.

In relativity theory, space and time become observer-dependent, i.e. movements of objects are no longer considered as properties of the objects, but only as relations in a frame of reference given by the observer, called an inertial frame. In quantum mechanics, we finally come to the realization that the observer not only influences what is observed, but even creates it.

The method of the exact sciences to find out the truth about their subjects of research is to carry out experiments and measurements. Scientific knowledge about the world always comes from interactions with the world. The focus is on interactions with the object of research with the help of tools, whereby "tool" must be understood in a very general way. A tool can also be a chemical compound that is brought together with another compound. If we want to measure something, we have to provoke a reaction from our object of research, because it is precisely the reproducible reaction to our provocation that convinces us that we have learned something about our object of research. This interaction is exact when we can observe the same reactions again and again.

To be able to achieve this exactness, we have to greatly simplify complex facts. We break down a composite object into its individual parts and we create constant framework conditions for our experiments. Depending on the experiment, this can be a constant room temperature, the avoidance of shocks, or any other factor. The exactness of science therefore always refers to artificial situations. Only that which is produced by humans in science can be produced exactly as desired. Nature itself is never exact. We only find exactness under special laboratory conditions.

But what knowledge do these exact interactions provide us with? *When* we measure something, we get a measured value. *When* we do an experiment, we *get* a result. We can then draw the conclusion that our object of research always reacts in the same way because it *has* a certain property, but this conclusion is only a possible interpretation and therefore neither logical nor exact. The only thing we have in the way of exact knowledge is the if-then relationship. A successful experiment provides as knowledge a guide to what we must do to get an exact response, and this means that scientific knowledge is nothing more than a guide to action. They tell us what we have to *do* to get an exact response.

And the same applies to our everyday perception. We may say the object is heavy even if we don't lift it, but in fact we only know that when we lift it, it resists and is perceived by us as heavy. The same happens when we attribute a property to our object of research. Even then, this attribution does not follow from the results of our experiments and measurements. It is always something that we add without the addition bringing us any additional knowledge.

Saying *the object is heavy* is shorter and less cumbersome than saying the object offers great resistance when we lift it. And of course, it is also much shorter and more elegant to say that our research object has this or that property, instead of tediously listing all the conditions of our experiment and describing how our research object reacts to them.

It is great that science finds and has found such guides to action. All our technical, economic, and social progress is based on these instructions. We can build machines of all kinds, we know which medicines produce which effects and which hygienic measures promote our health, we can produce artificial substances that have properties we desire, etc., etc. All these are true insights, but they always refer to actions.

Giambattista Vico was the first to formulate this as a general principle, saying that something can only be recognized as true if we have made it ourselves. While the scholastics (scholasticism is the philosophy of the Middle Ages) were of the opinion that truth lies in being, Vico says that truth lies in action.

Although scientific knowledge is always only a guide to action, all sciences continue to cling to the idea that the object of research has an existence of its own, is an entity, a something that can be grasped and held on to. Be it *the* society, *the* living being, *the* universe, or *the* elementary particle. We believe that our objects of research really do have these properties. Even without our activity. That therefore these properties have a being. We assign a thing to our action.

But as we have seen in sections 2.2 and 2.4, we are dealing with a lot of logical contradictions and paradoxes.

But what would change for us if we no longer sought truth in things but in action? Would we then live in a different world?

In fact, the world would not be a different world for us; everything would be exactly the same. The only difference would be that we would free ourselves from logical contradictions and paradoxes that we have brought upon ourselves. To return to the example of the heavy object: whenever we lift the object, we realize that it is heavy. And we do not add anything to this realization if we also assume that this heaviness is a property of the object. When we *do* something with the object, we know what kind of resistance we have to expect.

It is a little reminiscent of the problems that philosophy had with the almighty God. If this God is almighty, then He can create a stone that is so heavy that not even He himself can lift it. If this God is the creator of the world and the world develops all by itself afterwards, then such a God cannot be almighty.

And last but not least, the theological problem of theodicy should be cited, in which the legitimate question is asked how a benevolent and almighty God can allow so much suffering in the world. In the modern world, these problems have been solved simply by renouncing the idea of an almighty creator God.

And in the same way, we can free ourselves from the problems caused by the idea of a being that exists independently of perception, a *thing-in-itself.* Simply by renouncing this idea. The world in which we live, with all its appearances, would not be another world for us.

But we would not only be rid of logical contradictions and paradoxes. The nonsensical clinging to theories that block progress would no longer be possible. Deviations from if-then relationships would not be a problem at all. If we find a new method that helps us, we no longer have to fight against the prejudice of a scientific theory. If acupuncture helps us, then it is good. Whether acupuncture fits into some scientific theory is then no longer important. And we could justify our scientific actions differently.

According to Kant, science is not only based on observation, but also on asking questions of nature. Since asking questions always has a purpose, science is not free of purpose. And since nature does not answer, but scientists interpret the outcome of their experiments in a purposeful way, the answer is also not purpose-free.

Instead of continuing to look for the truth in things, the purpose of science could be to find functioning systems of action that serve to improve our quality of life. Not *what are* we looking for, but *why are* we looking?

## 2.8 Summary

Now that we have examined various terms used in physics, we have found a recurring pattern. We participate in the world by making observations, acting on the world, and changing it – and whenever we assume there is a being behind our actions, we run into problems with logic.

As soon as we assume that the world in which we live is something that also exists without us, we are dealing with paradoxes and contradictions.

The exact sciences are about objects and their properties. Physics is about the movement of objects in space and time. It is not by chance that we speak of the objective sciences, and by *objective* we always mean *real*. And we call a report *objective* if this report tells us how it *really* was. Objective properties in that sense are properties that really exist. We see an object and say it's there even when we're not looking. We lift an object, find it heavy, and say it would still have been heavy even if we hadn't lifted it. We measure the speed of an object and say it would still be moving even if we weren't measuring its speed.

But on closer inspection, all objective properties have turned out to be illusions. Objects have no properties except when we perceive them. Objects have no place and no speed unless we compare their movement with other movements. There are no objects without subjects, and whenever we think we have grasped something, it is grasping into the void.

Physics believes that it is a rational discipline and thus distinguishes itself from mystical and religious worldviews, because physics is about facts. But as we have seen on closer examination, this is a naïve and self-contradictory belief. Alfred North Whitehead writes about this:

> Science has never shaken off the impress of its origin in the historical revolt of the later Renaissance. It has remained predominantly an anti-rationalistic movement, based upon a naïve faith. What reasoning it has wanted, has been borrowed from mathematics which is a surviving relic of Greek rationalism, following the deductive method. Science repudiates philosophy. In other words, it has never cared to justify its faith or to explain its meanings; and has remained blandly indifferent to its refutation by Hume.

And this also applies to all sciences that have adopted the method of physics to gain true knowledge with the help of experiments and measurements.

# 3 Science and society

Even though science keeps telling us that what we used to believe - even though it was scientific knowledge - is no longer valid today, we do not doubt its competence. We shake our heads when we look at older scientific findings, that women are less intelligent than men, that butter is unhealthy and margarine and sugar are healthy, and so on. But we ignore the fact that probably in a hundred years people will shake their heads just as much about many of our scientific findings today.

So what is so fascinating about science that, despite all the changes in its findings, we cling to the belief that it can explain the world to us?

## 3.1 The success of the sciences

Aristotle gives less importance to the changeable than to the unchangeable. Phenomena and appearances of things change, but the underlying, the unchanging, is the nonesuch, the true, and the divine.

From this follows an image of society in which preservation has a higher value than change. In the Renaissance, this was reversed and dynamism became the highest principle. This dynamic generated the modern scientific worldview in which every change in a finding is celebrated as a success.

The success of the sciences is based on the fact that during the Renaissance, changing what existed was not seen as disruption but as progress. To this day, every scientist strives to replace existing knowledge with new knowledge. And the successful translation of new knowledge into successful action proves this attitude right.

Alfred North Whitehead, in his book *Science and the Modern World,* describes how science, which constantly changes its ideas and views, gains authority, while the Church loses authority every time it changes its views.

This results in a reassessment of values. It is no longer the unchanging that has the highest value, but the search for ever new insights. Darwin's theory of evolution and Einstein's theory of relativity are not seen as the defeat of science - although it had to give up its old ideas - but as triumphs.

Very often, the technical progress of the last few hundred years is seen as a success of science. In fact, technical and scientific development cross-fertilize each other and neither is the cause of the other. Rather, the cause lies in the spirit of optimism of the Renaissance, in which everything was questioned and the search for the new was assigned a higher priority than the preservation of the old. The loss of old truths opened up the search for new truths, and indeed the essence of progress lies in the conviction that any knowledge found and any technical achievement can be overcome and replaced by new knowledge and new technical achievements. In science, those who destroy existing truth and replace it with new truths are rewarded. One does not even have to find a new truth. Even the destruction of an existing truth is rewarded.

The fact that science and technology are not mutually dependent already follows from the fact that engineers have rarely been interested in the current state of basic research, and from the fact that technical developments endure even if science has found completely new explanations in the meantime. The construction of Cologne Cathedral is still a technical masterpiece, even though people back then had no idea about quantum physics and didn't even know the law of gravity and the law of inertia. Practical progress works according to trial and error, and the yardstick for success is not compatibility with the current state of science but usability, beauty, goodness, etc. – in short: quality, because every decision about accepting innovations is always a value judgement.

Scientific progress is also guided by values. The groundbreaking thing about Copernicus' idea was not that he replaced a wrong model with a correct one, because his model was also wrong and explained the celestial phenomena that could be observed worse than the Ptolemaic model with the Earth at the centre that was valid at the time. The really fascinating thing about Copernicus is that he removed the Earth from its static role as the centre of the universe and gave it motion.

Another reason for the success of science is that the machines that are built and laid out as analogues of the world or of man never fit exactly and therefore always have to be adapted and then supposedly always fit better. The improved machines also mean that the theory explaining the phenomena has to be constantly adapted. Behind this is the idea that this development towards ever new machines also represents an advance in knowledge. That not only are many problems of the living world solved, but that through this progress we also understand better and better why the world is the way it is, or appears to us the way it appears.

However, there are two decisive reasons for the worldwide triumph of science. One reason is the wonderful method of physics. You can learn it relatively easily and then get started without having to understand where this method comes from and why it works. And it is completely culturally neutral. All countries on Earth have successfully adopted this method. And the personal disposition of the people does not matter either. All progress would be inconceivable without this rational approach of soberly and systematically separating the wheat from the chaff in as unbiased a way as possible. It is the method of physics that helps us so successfully to solve problems and that turns a science into an exact science.

And the second decisive reason is the linking of the solution of problems with economic reward. Scientific knowledge is then no longer just a contribution in some library, but can become the starting point for founding a company that dramatically promotes the social and economic status of the discoverer.

But solving problems and explaining the world are not the same thing.

## 3.2 The ideal scientist and the method of physics

It is attributed as an advantage to science that the personality of the persons involved does not play a role. It is considered an advantage that scientists pursue their work neutrally and without prejudice and are only committed to the truth. Facts are facts and opinion is opinion, and science is about facts, not opinions.

Ortega y Gasset writes about this:

> When by 1890 a third generation assumes intellectual command in Europe we meet with a type of scientist unparalleled in history. He is one who, out of all that has to be known in order to be a man of judgment, is only acquainted with one science, and even of that one only knows the small corner in which he is an active investigator. He even proclaims it as a virtue that he takes no cognisance of what lies outside the narrow territory specially cultivated by himself, and gives the name of "dilettantism" to any curiosity for the general scheme of knowledge.
>
> What happens is that, enclosed within the narrow limits of his visual field, he does actually succeed in discovering new facts and advancing the progress of the science which he hardly knows, and incidentally the encyclopedia of thought of which he is conscientiously ignorant. How has such a thing been possible, how is it still possible? For it is necessary to insist upon this extraordinary but undeniable fact: experimental science has progressed thanks in great part to the work of men astoundingly mediocre, and even less than mediocre. That is to say, modern science, the root and symbol of our actual civilisation, finds a place for the intellectually commonplace man and allows him to work therein with success. The reason of this lies in what is at the same time the great advantage and the gravest peril of the new science, and of the civilisation directed and represented by it, namely, mechanisation. A fair amount of the things that have to be done in physics or in biology is mechanical work of the mind which can be done by anyone, or almost anyone. For the purpose of innumerable investigations it is possible to divide science into small sections, to enclose oneself in one of these, and to leave out of consideration all the rest. The solidity and exactitude of the methods allow of this temporary but quite real disarticulation of knowledge. The work is done under one of these methods as with a machine, and in order to obtain quite abundant results it is not even

necessary to have rigorous notions of their meaning and foundations.

These are strong words, but what of them?

In fact, a large part of scientific work is relatively mindless evaluation of data material or the systematic repetition of scientific experiments in which a parameter has been changed. All work that, in principle, machines could also do. Of course, people can do such work best if they approach it soberly and without prejudice. People who put aside their sexual inclinations, their political beliefs, their religious views, etc.

But should ideal scientists be like their object of research, namely a machine?

To what extent is someone who does this kind of work a scientist at all?

According to Kant, scientific work consists of asking questions of nature. But who asks the questions of nature and why does one ask questions of nature at all? Why should a person who has no personal qualities ask questions of nature? And what questions should he ask? Who decides which experiment to conduct, and who decides what the result means? Scientific progress always comes from creative minds that are able to interpret experimental results.

All genuine scientists who have been able to give their science a new direction or have founded their science in the first place absolutely do not correspond to this ideal image. Quite the opposite. Their religiosity, their interest in art, their involvement in social, political, or military interests are the driving forces behind their work.

The success of the physical method has led other sciences to adopt it. But when other sciences such as economics, biology, anthropology, etc. adopt the method of physics, they unfortunately forget that physics is machine theory. Conducting experiments with subsequent measurement or evaluating events with the aim of obtaining numbers that can be inserted into a mathematical law only leads to knowledge if the object of research is a machine. This is because the laws found describe a mechanism that is part of a machine. Although certainly not all people who work scientifically in economics, biology, anthropology, psychology believe that their object of research is a machine, they find mechanisms everywhere because of their approach. Market mechanisms, behavioural mechanisms, hereditary mechanisms, metabolic mechanisms, and so on and so forth. And that is of course understandable, because the point of the physical method is, after all, to search for mechanisms.

The beauty of the physics method is that it always produces results that can be marketed. You can learn this method without having to understand it. You apply it to any phenomenon and you always find some mathematical law. And since you have learned the method without understanding it, you also apply it to phenomena where the method does not fit at all. Because mathematical laws do not limit their own range of validity, scientists sometimes arrive at statements about extreme situations that are completely nonsensical. But not even in physics does this method yield any insights into what constitutes our life.

And although we believe that physics explains the world to us, it only knows how to successfully build machines.

## 3.3 The scientific view of the human being

We all know the phrases from machine theory when it comes to our state of mind. When we are angry, we need to let off steam, or we look for a valve. We go on a detox to regenerate our bodies, and when we're tired, it's time to recharge our batteries. When we need to remember something, we have it stored on our hard drive.

These metaphors also show quite clearly when they originated. Steam and valve come from the steam engine era already – it's not an exaggeration to call Freud's theory steam engine theory – while the hard drive was added in the computer age. Behind these metaphors is the idea that at least the body of a human is a machine. A highly complex machine, to be sure, but a machine nonetheless. We have René Descartes to thank for this view.

But it is not for nothing that the humanities came into being as a reaction to the neglect of everything that makes up the lifeworld of people – because people are not machines.

The success of physics in building machines seduces other sciences to adopt the method of physics. In doing so, they also adopt the belief that results of experiments and measurements have a higher value in their science than considerations. The idea of physics that experiments can provide true knowledge about the world and the belief that mathematics can help in this is uncritically adopted.

In economics, people believe that human behaviour can be described by simple mathematical equations, just like equations of motion in mechanics, and wonder why economics predictions are so often off the mark.

In brain research, there is the idea that the human brain is a computer, that there is such a thing as artificial intelligence and that in the future it will be possible to build computers that are capable of thinking like a human being. In the process, concepts of human thinking are often wrongly applied to machines. I would like to explain this using the example of a radiator thermostat.

A radiator thermostat is there to control the temperature. When it is too cold, the thermostat switches the heating on, when it is warm enough, it switches it off again. This is a simple mechanical regulator. But we could describe the thermostat in another way. We could say that when it is too cold, the thermostat *recognizes* that it is too cold, it *makes the decision* to turn on the radiator, and then *acts* according to its decision and turns on the radiator. It is immediately clear from this example that this is a nonsensical description of a simple mechanical device. But this is exactly what people do who claim that there are intelligent machines that can think. Yet the intelligence of a machine is the processing of if-then decisions that have been programmed. Apparently, not everyone in brain research is aware that thinking includes speaking and that thinking, just like speaking, is a social phenomenon. Therefore, neither an isolated brain nor a computer can think.

Medicine has also taken the machine view of humans very far. Modern medicine has become purely a repair business, as if people had to be repaired. Even though there are many people in the healthcare system who care very much about the well-being of patients, medicine sees its task as making a sick person healthy. And *healthy* means *functional again*. It does this in the same way as a car mechanic repairs a car. A sick person is treated like a broken car. Just as a car mechanic checks the mechanical parts for functionality and tests the levels and consistencies of fluids, a doctor checks the mobility of the body, tests the concentration of various substances in the body fluids, and, possibly, detects deviations from target values. If a car mechanic locates the problem in the starter motor or in the gearbox or in the alternator, then he only deals with the starter motor, the gearbox, or the alternator. And that's how a doctor works. If he locates the problem in the kidney or in the intestine or in the brain, he takes care of the kidney, the intestine, or the brain.

If all tested parameters are within the target range, then a person is considered healthy, no matter how they feel; and if there are deviations, then the patient is advised to seek treatment, even if the patient feels well.

There are of course many parts of the human body that have similarities to machines and for which this procedure can be successfully applied. But there are also diseases that do not fit into the concept of a functioning machine, which explains the existence of so many alternative healing methods. It would be desirable for us to look at alternative healing methods not only from the point of view of whether or not they can be integrated into the machine model.

It would also be nice if science finally took note of the fact that people are not machines and brains are not computers; that people are creative and can think for themselves.

It is the human being who created science. It is man who made the world we live in. Everything that experimental science finds was created by them, and then they pretend they weren't even there.

In scientific experiments with the inanimate world, this error in thinking does not lead to findings about the world itself, but at least one arrives at working descriptions of the world created by humans. What has been created by science is then not the result of a discovery process, but it is there nonetheless and can be used for practical purposes.

The situation is completely different when scientists deal with people. Then there is no difference between the object of research and the observer. Both are human beings. Experimental scientific work in psychology is often an arranged experiment and consequently an artificial situation. But the scientist, when describing the experiment, simply remove themselves from the picture and claim that they have discovered a characteristic of the subjects that has nothing to do with either the prior arrangement or the artificial situation of the experiment.

However, a scientific experiment with humans leads at best to a description of human behaviour in experimental situations. An extension of this description to situations outside the experimental situation is purely arbitrary and does not lead to an explanation of humans or their behaviour.

## 3.4 Our world is a man-made world

Since humans have become sedentary, their environment increasingly consists of things they have made themselves, because one human skill is making tools and creating buildings and machines. The world we live in allows this to happen. It provides the material, but people have to have needs that make them create. In creating this reality, people understand that the environment offers resistance, which is often not arbitrary, but obeys certain laws. In order to build a house, static properties of the building materials have to be taken into account. This means that people evaluate their activities and learn from failures. This distinguishes them from the buildings and machines they create, since in the best case buildings or machines work, or otherwise do not work.

Humans can change their environment through creative acts and thus create their living world. Therefore, the idea that every change in the world comes about in the same way as human-made changes is obvious: that natural disasters, for example, are caused by angry gods. And ideas arise that the world as a whole is a created world. Just as humans created their reality, someone or something created the whole world.

Access to the world with the help of tools, which makes it possible to change the world through buildings and machines, is access to the world against the world. *Subdue the world*. The view of philosophy and science arises from this access to the world.

But subduing the world and eating from the tree of knowledge are not the same thing. We can ask what possibilities there are to change the world and according to what laws the world can be changed. But the answer does not give us any conclusions as to why the world is the way it is.

Everything we know about the world quite precisely through philosophy and science is based on knowledge about created worlds. The model for Plato's reflections is the man-made building and for Galileo and Descartes it is the man-made machine. Plato divided the world into a world of ideas and a world of matter. In ancient Greece, the term *idea* referred to an architect's plan for a building, which could be discussed before the building was erected. Descartes split the world into a world of machines and a world of mind.

Both philosophy and physics are about created worlds, and therefore both only work if there is a creator of the world. And just as a building or a machine is created and then exists, one can imagine that the world was created at some point and has existed ever since; that is, that the world has a beginning.

But if today we rightly believe that the creator of the world is not God but the autonomous human being, then the world can neither be a machine nor have a beginning.

Everything that fact-based science does and says presupposes the existence of humans. Only humans are capable of building machines, conducting experiments, taking measurements, creating facts. Experimental science is the failed attempt to observe a world created by humans and pretend that this world would also exist without humans.

# Final remarks

Physics tells us that we live in a cold, inhospitable world. On the edge of an insignificant Milky Way somewhere in the universe. That reality consists exclusively of elementary particles moving in space and time. And that much of what makes up our living world, namely compassion, love, suffering, happiness, and death, are only subjective sensations that do not exist in reality.

The American philosopher E. A. Burtt writes about this:

> The gloriously romantic universe of Dante and Milton, that set no bounds to the imagination of man as it played over space and time, had now been swept away. Space was identified with the realm of geometry, time with the continuity of number. The world that people had thought themselves living in - a world rich with colour and sound, redolent with fragrance, filed with gladness, love and beauty, speaking everywhere of purposive harmony and creative ideals - was crowded now into minute corners in the brains of scattered organic beings. The really important world outside was a world hard, cold, colourless, silent, and dead; a world of quantity, a world of mathematically computable motions in mechanical regularity.

But the worldview of physics, according to which only what we call objective really exists, is a naïve belief that has failed because of its own contradictions.

Galileo's and Newton's machine model with its mechanics and the associated mechanisms cannot explain the world to us, even if physics does not want to admit this. Peter Sloterdijk has found a fitting formulation for this:

> Like all institutions imbued with the will to survive, they knew how to weather the "dissolution" of their foundations.

To avoid any misunderstandings, this book is not about declaring war on rationalism. Quite the opposite. The greatest achievement of humanity is enlightened rationalism, which has brought us human rights, freedom of expression, the rule of law, and, last but not least, our prosperity.

But the path taken by Galileo, Descartes, and Newton and supported by Kant is not the path of the Enlightenment and rational reason. That is why it is no coincidence that after the increasing recognition of physics, the humanities emerged, which, at least in the beginning, tried to salvage what could still be salvaged from the rational Enlightenment.

Nor is it about questioning scientific achievements that have significantly improved our quality of life, such as the successful fight against hunger and disease.

It is exclusively about ending the interpretative sovereignty of the so-called exact sciences that want to explain the world to us. What does it mean for a democratic state if there is a non-democratic institution like science that is granted an authority that is not controlled by independent bodies? It would be as if we were living in a state of God that is officially a democracy but in which, in case of doubt, the Church has the last word.

And what authority can an institution have that is based on a faith that cannot be shaken by anything and that calls all perceptions of people that do not fit into its belief system subjective imagination and consequently not real? While other belief systems at least give people hope and comfort, physics wants us to believe that the world is a senseless, cold, and inhuman place.

If we want to continue enlightened rationalism and keep reason as the highest standard for gaining knowledge, then we have to get away from the idea that physics and its method of experimentation and measurement can help us do that. Fundamental physics is not a rational discipline, and therefore we must protect reason from it. We must free people's thinking from a one-sided focus on everything measurable. What is measurable is not all that is real and determines our lives. We must not leave the sciences to the machine people who, in biology, seek life in the machine model of the cell and, in brain research, seek the spirit and consciousness in the neurological linking of brain cells. Machines and technology are good, mathematics and measurements are helpful, but they are only part of what makes up our living world.

Existential questions that determine our lives and cannot be pressed into an ideology of objective and measurable reality must not be left to people who measure instead of thinking.

The words of Odo Marquard illustrate this very well:

> Where is the soul in the empirical science of psychology and the spirit in the cognitive sciences of artificial intelligence? ... Where is the human living world in astrophysical cosmology, in which we humans have to live through and die through our lives every day? Where is the subject of God, freedom, meaning in the hard sciences? Where is the individual, the traditions, the stories in evolutionary biology? [Translation by Calwius]
> (Wo bleibt in der empirischen Wissenschaft Psychologie die Seele und in den Kognitionswissenschaften der künstlichen Intelligenz der Geist? … Wo bleibt in der astrophysikalischen Kosmologie die menschliche Lebenswelt, in der wir Menschen tagtäglich unser Leben durchleben und durchsterben müssen? Wo bleibt in den harten Wissenschaften das Thema Gott, Freiheit, Sinn? Wo bleiben in der Evolutionsbiologie das Individuum, die Traditionen, die Geschichten?)

In demystifying the world through science, we have thrown the baby out with the bathwater. If superstition and irrationality rightly disappear, then it does not follow that we should view the world like a machine. Then it does not follow that everything is predetermined, because there is undeniably human creativity that can create something new. Living beings can distinguish between good and bad, machines not even between right and wrong.

What possibilities for reorientation present themselves when we stop believing that the sciences can explain the world to us? Do we want to look back at ancient wisdom teachings and try to incorporate them into our modern world? Or do we want to try to find new ways? And what might a new way look like?

Indian philosophy has come just as far rationally as Western philosophy. But they stopped at the realization that an objective worldview is not true, but only functions. In the West, on the other hand, this functioning was seen as an incentive to gain ever new functioning insights. Even if it is true that this Western approach has not yet brought fundamental insights into people's existential questions, it has nevertheless dramatically improved people's living conditions: human rights, the rule of law, and prosperity.

A new way could be to free ourselves from the false idea that only what can be measured exists. Then we are free for new insights into existential questions. Then we can do again what was done at the beginning of the Enlightenment, namely assess trial and error rationally. Not what is measurable is the standard, but what works.

For example, a scientific explanation of why a certain medicine has a certain effect is neither true nor does it contribute anything to the effect of that medicine. What counts is the description of how the medicine changes the way we feel. Therefore, it does not matter whether a scientific explanation is found for a treatment method, such as homeopathy or acupuncture. As long as it helps people, it is good.

So we should continue on the path of rational enlightenment and just say goodbye to the idea that there is such a thing as objective truths without humans involved. The attempt of physics and the exact sciences to find a world outside of man has failed. This means that man is the measure of all things because there is no other measure at all.

Only humans are capable of developing ideas about the world. Our nature and our blue planet only exist for humans. However, it does not necessarily follow from this that nature exists for humans or that humans have to subjugate the Earth.

After all, there is a huge difference between the past and the present in terms of the possibilities of a secure existence for all people. No human being would have to wage a merciless struggle for existence just to survive. There is also a world approach that is in harmony with the world. In a world situation where the survival of humanity could be ensured without problems, we can afford to take a less confrontational approach to the world. Instead of using nature as a means to the end of knowledge or profit maximization, we could use our technical abilities exclusively for the benefit of humanity.

There is also no reason to assume that the current problems of humanity, such as the pollution of our planet, the extinction of species, and climate change, cannot be solved through a rational approach.

It will be crucial that we get into a spirit of optimism that encourages people to tackle the problems at hand and not to despair from self-accusation about the past that brought us these problems. Norbert Bolz came up with the beautiful sentence

> We are not afraid of the abyss, we build a bridge over it. [Translation by Calwius]
> (Wir haben keine Angst vor dem Abgrund, wir bauen eine Brücke darüber.)

What might a new view of the world look like if we abandon the idea that there are such things as objective facts?

Instead of blindly trusting in science to tell us what is right and what is wrong, we could orient our lives according to what is good and what is bad for us, as Odo Marquard rightly said:

> The world of reflection and the lifeworld must not become unrelated to each other. [Translation by Calwius]
> (Reflexionswelt und Lebenswelt dürfen nicht beziehungslos zueinander werden.)

If we no longer look for absolute truths, we no longer need guides to tell us what is right and wrong. Even if we follow Kant's call

> Have the courage to use your own understanding

we create a more liveable world.

Already 2,500 years ago, Siddhartha Gautama, the historical Buddha, got to the heart of the matter:

> Do not go upon what has been acquired by repeated hearing; nor upon tradition; nor upon rumour; nor upon what is in a scripture; nor upon surmise; nor upon an axiom; nor upon specious reasoning; nor upon a bias towards a notion that has been pondered over; nor

> upon another's seeming ability; nor upon the consideration, "The monk is our teacher."
>
> When you yourselves know: "These things are bad; these things are blameable; these things are censured by the wise; undertaken and observed, these things lead to harm and ill", abandon them.
>
> When you yourselves know: "These things are good; these things are not blameable; these things are praised by the wise; undertaken and observed, these things lead to benefit and happiness", enter on and abide in them.

But obviously people were not yet ready to understand this message. What emerged after Buddha under the name *Buddhism* has little to do with it. Religious leaders like the Dalai Lama in Tibetan Buddhism are certainly the opposite of what Buddha wanted.

But perhaps by now we have come to understand this message and are ready to take our destiny into our own hands.

We should be open to ways of thinking that are not oriented towards mathematics and the machine model. People can perceive much more than what we can measure. What is true is not only what science can measure.

This requires enlightened people who participate in society, because experts are not only scientists, but all people who know something about their work. Everything that regulates our coexistence should be worked out in democratic processes.

Science should be fundamentally integrated into the social discussion. Scientists should seek social justification for their actions. Since science cannot gain knowledge about the world per se, but only functions as a problem solver, science should solve problems that have something to do with the people who make its research possible.

Exact science, which believes it can gain all knowledge through experiments and measurements, should receive its justification from those humanities that still place the primacy of thought before experiment.

We humans have an insatiable longing for an explanation of our existential situation, and since the sciences with their experiments and objective measurements cannot offer us any explanations, we have to look for alternatives. Philosophy and the humanities can help us, provided they refrain from gaining objective knowledge through measurements. And it would be helpful if the sciences took note of the fact that the good and the beautiful have always been the impetus and yardstick for scientific action.

# Notes and references

## Introduction

The quote from Peter Sloterdijk is taken from his book *Critique of Cynical Reason* (page 87).

Kant says about science that only the idea of addressing questions posed by reason to nature, and not simply observing it, makes empiricism a science. This is not a literal quotation, but it can be found in his 2nd Preface of *The Critique of Pure Reason*.

On the idea that the world is a machine, E. A. Burtt writes in *The Metaphysical Foundation of Modern Science*:

> Galileo's exact study of velocities and accelerations forced him to devise a simple technique for the geometrical representation of time, which was fairly adequate to the truths he sought to illustrate. With him, the physical world begins to be conceived as a perfect machine whose future happenings can be fully predicted and controlled.
>
> The conception of the world as a perfect machine is thus rendered inevitable, and it is no accident that first in Huyghens and (in a more philosophical form) in Leibniz we have this position unequivocally proclaimed.
>
> Such, then, is Descartes' famous dualism – one world consisting of a huge, mathematical machine, extended in space; and another world consisting of unextended, thinking spirits.

Karl-Heinz Brodbeck also writes on the subject of the machine in his book *Die fragwürdigen Grundlagen der Ökonomie* (The questionable foundations of economics) on page 37:

> The historical model of all mechanics is the machine, and its prototype in turn is the clockwork. Ptolemaic astronomy had already used mechanical models of the planets (planetaria) early on, which then became the model for clockworks. The mechanical meshing of gears in turn became a model of thought for the entire natural world. It was above all Descartes who "described the Earth and the whole visible world in the manner of a machine", but Descartes' critic Leibniz

also says: "What I maintain is that the clockwork of the world, without needing any adjustment, goes on." [Translation by Calwius] (Das historische Vorbild aller Mechanik ist die Maschine, und deren Prototyp wiederum ist das Uhrwerk. Die ptolemäische Astronomie hatte bereits früh mechanische Modelle der Planeten (Planetarien) verwendet, die dann zum Vorbild der Uhrwerke wurden. Das mechanische Ineinandergreifen der Zahnräder wurde seinerseits zu einem Denkmodell für die gesamte natürliche Welt. Es war vor allem Descartes, der „die Erde und die ganze sichtbare Welt nach Art einer Maschine beschrieben" hat, aber auch Descartes Kritiker Leibniz sagt: „Was ich behaupte, ist, daß das Uhrwerk der Welt, ohne einer Nachbesserung zu bedürfen, fortgeht.")

## 1 Physics' method of understanding the world

On the thesis that physics has shifted the focus from *why* to *how,* Alfred North Whitehead writes in *Science and the Modern World*:

Galileo keeps harping on how things happen, whereas his adversaries had a complete theory as to why things happen.

A nice explanation of what induction is has been given by Robert Pirsig in his book *Zen and the Art of Motorcycle Maintenance*:

Two kinds of logic are used, inductive and deductive. Inductive inferences start with observations of the machine and arrive at general conclusions. For example, if the cycle goes over a bump and the engine misfires, and then goes over another bump and the engine misfires, and the goes over another bump and the engine misfires, and then goes over a long smooth stretch of road, and there is no misfiring, and then goes over a fourth bump and the engine misfires again, one can logically conclude that the misfiring is caused by the bumps. That is induction: reasoning from particular experiences to general truths.

Deductive inferences do the reverse. They start with general knowledge and predict a specific observation. For example, if, from reading the hierarchy of facts about the machine, the mechanics knows the horn of the cycle is powered exclusively by electricity from the battery, then he can logically infer that if the battery is dead the horn will not work. That is deduction.

On mathematics in physics, Galileo says in *Il Saggiatore*:

The book of nature is written in the language of mathematics [...] .

## 2.1 Insights from machine theory

On the question of whether the *world* machine needs maintenance, E. A. Burtt writes in *The Metaphysical Foundation of Modern Science*:

God, now the chief mechanic of the universe, has become the cosmic conservative. His aim is to maintain the status quo. The day of novelty is all in the past; there is no further advance in time. Periodic reformation when necessary, by the addition of the indicated masses at the points of space required, but no new creative activity - to this routine of temporal housekeeping is the Deity at present confined.

Historically, the Newtonian attempt thus to keep God on duty was of the very deepest import. It proved a veritable boomerang to his cherished philosophy of religion, that as the result of all his pious ransackings the main providential function he could attribute to the Deity was this cosmic plumbery, this meticulous defence of his arbitrarily imposed mechanical laws against the threatening encroachments of irregularity. (...)

But science moved on, and under the guidance of the less pious but more fruitful hypothesis that it would be possible to extend the mechanical idea over an ever wider realm, Newton's successors accounted one by one for the irregularities that to his mind had appeared essential and increasing if the machine were left to itself.

The following three examples refer to doubts about the dogma machine:

Israel Rosenfield in *The Strange, Familiar and Forgotten*, p. 7

The continuity of consciousness derives from the correspondence which the brain establishes from moment to moment. Without this activity of connecting, we would merely perceive a sequence of unrelated stimuli from moment to unrelated moment, and we would be unable to transform this experience into knowledge and understanding of the world. This is why conscious human knowledge is so different from the *knowledge* that can be stored in a machine or in a computer.

Whitehead, Alfred North: *Science and the Modern World*, p. 130.

Evolution, on the materialistic theory, is reduced to the rôle of being another word for the description of the changes of the external relations between portions of matter. There is nothing to evolve, because one set of external relations is as good as any other set of external relations. There can merely be change, purposeless and unprogressive.

Daniel J. Nicholson has written an article on the subject, entitled *Is the Cell Really a Machine?*

### 2.2.1 Objects

But here we have a logical contradiction, because if it is not recognizable, then we cannot recognize that it is outside our cognition...: Karl-Heinz Brodbeck writes about this in *Zirkel des Wissens* (The circle of knowledge):

Cognition takes place in something from which one cannot step out. One cannot say that there is an "outside" for cognition that is not cognition. That would simply be a logical contradiction. First one asserts, "I cognize something as existing outside cognition"; but questioned more closely, one replies, "It is not cognizable." Thus, cognition is a modification in a wherein, not the emergence of something from its opposite (an unknowable thing-in-itself). [Translation by Calwius]

(Das Erkennen vollzieht sich in etwas, aus dem man nicht heraustreten kann. Man kann nicht sagen, es gäbe ein ‚Draußen' für die Erkenntnis, das keine Erkenntnis ist. Das wäre schlichtweg ein logischer Widerspruch. Zuerst behauptet man: „Ich erkenne etwas als außerhalb der Erkenntnis Existierendes"; doch näher befragt, antwortet man: „Es ist nicht erkennbar." Das Erkennen ist also eine Modifikation in einem Worin, nicht die Entstehung von etwas aus seinem Gegenteil (einem unerkennbaren Ding an sich).)

Consequently, we cannot avoid the realization that whenever we say that something *is,* that then also means that it is recognized...: Once again Karl-Heinz Brodbeck in *Zirkel des Wissens* (The circle of knowledge):

> Only in cognitive space do things "exist" as recognized things. Whatever *is,* always has at the same time the meaning of being known. It is impossible to speak of a being of which there is no knowledge. For even the "great unknown", the "thing-in-itself", the "unknowable" is related to cognition precisely because we determine it through a negation. The negation of "something" is defined as a non-thing by the something. [Translation by Calwius]
> (Nur im kognitiven Raum ‚existieren' Dinge als erkannte Dinge. Was immer *ist,* hat auch stets zugleich die Bedeutung, erkannt zu sein. Von einem Seienden zu sprechen, von dem es keine Kenntnis gibt, ist unmöglich. Denn selbst das ‚große Unbekannte', das ‚Ding an sich', das ‚Unerkennbare' ist eben dadurch, daß wir es durch eine Negation bestimmen, auf die Erkenntnis bezogen. Die Negation von ‚Etwas' ist als ein Nicht-Etwas durch das Etwas definiert.)

(...) there is no perceiver without the perceived and no perceived without the perceiver. Both always arise simultaneously and do not exist without the other...: Finally again Karl-Heinz Brodbeck in *Zirkel des Wissens* (The circle of knowledge):

> Subject is that which is cognizantly related to an object. If a subject does not relate to an object, then this is a nonsensical concept. Although "relating" may always come up empty, like – to use another metaphor – the grasping hand, it remains preserved as a relation. Occam's philosophy and later phenomenology call this "intentionality" – that is, being directed towards something. In the hopeful expectation of an object, the subject is always related to an object. And the reverse is true for the object: it is the what-to of intentionality. The logic of this relationship actually becomes very clear in the father–child metaphor. Here, the possible errors in thinking are also very easy to recognize. Heraclitus says that it is only the child that makes the father a father. This is true, but of course it is also true the other way round. Someone can only be called "father" if he has or had a child. [Translation by Calwius]

(Subjekt ist das, was erkennend in Beziehung zu einem Objekt steht. Bezieht sich ein Subjekt nicht auf ein Objekt, dann ist das ein unsinniger Begriff. Zwar mag das ‚Beziehen' immer wieder leer ausgehen, wie – um eine andere Metapher zu verwenden – die ergreifende Hand; es bleibt doch als Relation bewahrt. Die Philosophie Occams und später die Phänomenologie nennen dies ‚Intentionalität', das heißt auf etwas ausgerichtet zu sein. In der hoffenden Erwartung eines Objekts ist das Subjekt immer auf ein Objekt bezogen. Und für das Objekt gilt das Umgekehrte: Es ist das Worauf der Intentionalität.

Die Logik dieser Beziehung wird tatsächlich in der Vater-Kind-Metapher sehr deutlich. Hier sind auch die möglichen Denkfehler sehr einfach zu erkennen. Heraklit sagt, daß es erst das Kind ist, das den Vater zum Vater macht. Das ist richtig, gilt aber natürlich auch umgekehrt. Jemand kann nur ‚Vater' genannt werden, wenn er ein Kind hat oder hatte.)

### 2.2.2 Movement

On the subject of the EPR paradox, see: Hoffmann, Martin: *A Local Realistic Explanation of EPR. Correlations,* in Foundations of Physics, Vol. 20, No. 8, 1990; Hoffmann, Martin: *Fernkorrelationen in der Quantentheorie. Eine neue Interpretation (Long-distance correlations in quantum theory: A new interpretation),* Doctoral thesis, University of Tübingen, 1988

### 2.3 The observer

On the explicit introduction of the term "observer" into physics:

> If an observer be stationed at A with a clock, he can estimate the time of events occurring in the immediate neighbourhood of A, by looking for the position of the hands of the clock, which are synchronous with the event. If an observer be stationed at B with a clock,—we should add that the clock is of the same nature as the one at A,—he can estimate the time of events occurring about B

is a quote by Einstein, Albert: *The Principle of Relativity,* Library of Alexandria, 2022

I have taken the reflections on the subject of the observer and the observed from the book *Der Zirkel des Wissens* (The circle of knowledge) by Karl-Heinz Brodbeck.

### 2.4 Cause and effect

So we have the experience that there is something we call cause and something we call effect, but without this experience we cannot discover the effect in the cause: David Hume writes about this in his book *An Inquiry into the Human Mind*:

> In a word, then, every effect is a distinct event from its cause. It could not, therefore, be discovered in the cause, and the first invention or conception of it, a priori, must be entirely arbitrary.

The quote

> If things did not exist Without essence, The phrase, "When this exists so this will be," Would not be acceptable.

is found in Garfield, Jay L.: *The Fundamental Wisdom of the Middle Way, Nagarjuna's Mulamadhyamakakarika,* 1.10

So in order to continue to speak rightly of cause and effect, we must detach ourselves from the idea that cause and effect are each separate entities...: Karl-Heinz Brodbeck writes about this in his book *Zirkel des Wissens* (The circle of knowledge) on p. 221:

> The cause is to bring forth the effect. This bringing forth can only happen in such a way that the cause ceases to be distinct from the effect; "if" the cause is an entity that exists for itself, it is the entity of non-effect. How can a non-effect bring forth an effect, how can nothingness bring forth being? To be the cause of an effect, the cause must have no being for itself; the same is symmetrically true of the effect. It is like father and child. One cannot speak of the "being" of a father for himself without a child. [Translation by Calwius]
> (Die Ursache soll die Wirkung hervorbringen. Dieses Hervorbringen kann nur so geschehen, daß die Ursache aufhört, sich von der Wirkung zu unterscheiden; ‚ist' die Ursache eine für sich seiende Entität, so ist sie die Entität der Nicht-Wirkung. Wie kann eine Nicht-

Wirkung eine Wirkung hervorbringen, wie kann das Nichts das Sein hervorbringen? Um Ursache einer Wirkung zu sein, darf die Ursache kein Sein für sich haben; dasselbe gilt symmetrisch für die Wirkung. Es ist wie bei Vater und Kind. Man kann nicht vom ‚Sein' eines Vaters für sich sprechen, ohne Kind.)

## 2.5 Induction

On the subject of induction, David Hume writes in *An Inquiry into the Human Mind*:

It is impossible, therefore, that any arguments from experience can prove this resemblance of the past to the future; since all these arguments are founded on the supposition of that resemblance.

## 2.7 Is the truth in the action or in the things?

While the scholastics (scholasticism is the philosophy of the Middle Ages) were of the opinion that truth lies in being, Vico says that truth lies in action:

The criterion of what is true is the doing itself.

is a quote by Vico, Giambattista: *Liber metaphysicus*

## 2.8 Summary

The quotation from Alfred North Whitehead comes from his book *Science and the Modern World.*

## 3.1 The success of the sciences

(...) the unchanging is the highest, the true, and the divine: Odo Marquard writes on this in *Universalgeschichte und Multiversalgeschichte* (Universal history and multiversal history) in: *Zukunft braucht Herkunft* (Future needs heritage), p. 110:

(...) in pre-modern times, ontological credit was traditionally only given to that which is unchangeable: that which remains unchangeably the same; and changeability was regarded as an ontological evil (...) [Translation by Calwius]

((...) ontologische Bonität hatte vormodern traditionell nur das Unveränderliche: das, was unwandelbar sich selber gleich bleibt; und die Wandelbarkeit galt als ontologisches Übel (...))

This dynamic generates the modern scientific worldview, in which every change in a finding is celebrated as a success...: Odo Marquard again in *Theodizeemotive in der neuzeitlichen Philosophie* (Theodicy motifs in modern philosophy) in: *Zukunft braucht Herkunft* (Future needs heritage), p. 137:

(...) curiosity is transformed from a vice into a central scientific virtue; and it is above all the error that is disenchanted. In modern times, it has a steep positive career and finally - as a productive fiction - moves into the position of the most important conditions for knowledge and action; as a "lie in the extra-moral sense" and "expedient error" (...) with Popper - the history of science: *we err upwards.* [Translation by Calwius]

((...) die Neugier wird aus einem Laster zur zentralen Wissenschaftstugend; und entübelt wird vor allem der Irrtum. Er macht modern eine steile Positivkarriere und rückt schließlich - als produktive Fiktion - ein in die Position der wichtigsten Erkenntnis - und Handlungsbedingungen; als ‚Lüge im außermoralischen Sinn' und ‚zweckmäßiger Irrtum' (...) bei Popper - die Wissenschaftsgeschichte: *Wir irren uns empor.)*

Alfred North Whitehead, in his book *Science and the Modern World*, describes how science, which constantly changes its ideas and views, gains authority, while the Church loses authority every time it changes its views:

In the first place for over two centuries religion has been on the defensive, and on a weak defensive. The period has been one of unprecedented intellectual progress. In this way a series of novel situations have been produced for thought. Each such occasion has found the religious thinkers unprepared. Something, which has been proclaimed to be vital, has finally, after struggle, distress, and anathema, been modified and otherwise interpreted. The next generation of religious apologists then congratulates the religious world on the deeper insight which has been gained. The result of the continued repetition of this undignified retreat, during many generations, has at last almost entirely destroyed the intellectual authority of religious thinkers. Consider this contrast: when Darwin or Einstein proclaim theories which modify our ideas, it is a triumph for science. We do not go about saying that there is another defeat for science, because its old ideas have been abandoned.

It is the method of physics that helps us so successfully to solve problems and that turns a science into an exact science. Odo Marquard writes about this in *Zukunft braucht Herkunft* (Future needs heritage), p. 264:

The modern - exact - natural sciences, in order to be exact, must make reality laboratory-ready and their scientists interchangeable, and in return neutralize the histories that make up the human living world. For the exact scientist, it must be indifferent from which - linguistic, religious, cultural, familial - traditions he comes; only in this way - interchangeable, because history-neutral - can he arrive at world-unifying verifiable results by measuring and experimenting. [Translation by Calwius]
(Die modernen - exakten - Naturwissenschaften müssen, um exakt zu sein, die Wirklichkeit laborfähig und ihre Wissenschaftler austauschbar machen und dafür die Geschichten neutralisieren, aus denen die menschliche Lebenswelt besteht. Für den exakten Wissenschaftler muss es gleichgültig sein, aus welchen - sprachlichen, religiösen, kulturellen, familiären - Traditionen er kommt; nur so - austauschbar, weil geschichtsneutral - kann er messend und experimentierend zu welteinheitlich nachprüfbaren Resultaten kommen.)

And the second reason is the linking of the solution of problems with economic reward: Norbert Bolz writes about this in his book *Der alte weiße Mann* (The old white man):

Of course, there have been great discoveries, inventions, and scientific findings all over the world, especially in India, China, and the Arab countries. But only the Occident, the West, has placed economic premiums on their technical use. Added to this is the organizing power of a rational, i.e. calculable, law, which has been cultivated by trained jurists in the tradition of Roman law. (...) The West explored ever new possibilities of application. The Chinese, Hindus, and Muslims, on the other hand, lacked the interest in exploiting their technologies and lacked the capitalist profit motive. Western culture rewards the technical use of scientific knowledge. So it is not science that is specifically European, but the way it is put to use.

Modern capitalism places premiums on the technical application and economic exploitability of science. [Translation by Calwius] (Natürlich hat es überall auf der Welt, vor allem in Indien, China und den arabischen Ländern, großartige Entdeckungen, Erfindungen und wissenschaftliche Erkenntnisse gegeben. Aber nur der Okzident, das Abendland, der Westen hat ökonomische Prämien auf ihre technische Verwendung gesetzt. Hinzu kommt die Ordnungsleistung eines rationalen, also berechenbaren Rechts, das von geschulten Juristen in der Tradition des Römischen Rechts gepflegt worden ist. (...) Der Westen erkundete immer neue Anwendungsmöglichkeiten. Bei den Chinesen, Hindus und Moslems dagegen fehlte das Interesse an einer Ausbeutung ihrer Technologien und es fehlte das kapitalistische Profitmotiv. Die abendländische Kultur belohnt die technische Verwendung wissenschaftlicher Erkenntnisse. Spezifisch europäisch ist also nicht die Wissenschaft, sondern die Art ihrer Indienstnahme. Der moderne Kapitalismus setzt Prämien auf die technische Anwendung und wirtschaftliche Verwertbarkeit von Wissenschaft.)

## 3.2 The ideal scientist and the method of physics

The quote from Ortega y Gasset comes from his book *The Revolt of the Masses*, p. 110.

On the role of the scientist, Albert Einstein writes in *The Evolution of Physics* on p. 5:

To obtain even a partial solution the scientist must collect the unordered facts available and make them coherent and understandable by creative thought.

And again on p. 33:

Physical concepts are free creations of the human mind, and are not, however it may seem, uniquely determined by the external world.

But not even in physics does this method lead to any insights into what constitutes our life. Peter Sloterdijk writes about this in his book *Critique of Cynical Reason,* p. 179:

The principle of experience, in the last analysis, bursts all moralism, including the scientific method. What life is grasped by the researcher not in the theoretical attitude but solely through the leap into life itself. (...)

Scientists who hide their will to power from themselves and conceive of experience only as a knowledge about "objects" *cannot* achieve that knowledge acquired by accumulating experience in the form of a journey to the "real" things. For the empirical amoralist, life is not an object but a medium, a journey a practical essay, a project of alert existence.

### 3.3 The scientific view of the human being

But it is not for nothing that the humanities came into being as a reaction to the neglect of everything that constitutes the lifeworld of human beings – because human beings are not machines...: See Marquard, Odo: *Über die Unvermeidlichkeit der Geisteswissenschaften* (On the inevitability of the humanities), in: *Zukunft braucht Herkunft* (Future needs heritage), p. 171ff.

Karl-Heinz Brodbeck explains why economic predictions are so often off the mark in *Die fragwürdigen Grundlagen der Ökonomie* (The questionable foundations of economics), section 2.2.

### 3.4 Our world is a man-made world

This approach to the world gives rise to the view of philosophy and science. Karl-Heinz Brodbeck writes about this in *Handeln als metaphysischer Horizont* (Action as a metaphysical horizon):

> What is true is no longer what reason abstracts from active nature; only what man himself does is recognized as true. A technical concept of truth emerges that places the experimenter as the successor of the craftsman. [Translation by Calwius]
> (Wahr ist nicht mehr, was eine Vernunft erkennend aus der wirkend-tätigen Natur abstrahiert, als wahr wird nur noch das erkannt, was der Mensch selbst macht. Es entsteht ein technischer Wahrheitsbegriff, der den Experimentator als Nachfolger des Handwerkers einsetzt.)

But if we rightly believe today that God is not the creator of the world, but autonomous man, then the world can neither be a machine nor have a beginning: See Marquard, Odo: *Theodizeemotive in der neuzeitlichen Philosophie* (Theodicy motifs in modern philosophy) in: *Zukunft braucht Herkunft* (Future needs heritage), p. 133.

(...) God *has* left creating, because it is not God who is the creator of the world, but - autonomistically - man, and specifically - according to Kant - as the creator of the artificial experimental world of the exact sciences and their technical world of application (...) [Translation by Calwius]
((…) Gott *hat* das Schaffen bleibenlassen, denn nicht Gott ist der Schöpfer der Welt, sondern - autonomistisch - der Mensch, und zwar - so Kant - als Schöpfer der artifiziellen Experimentalwelt der exakten Wissenschaften und ihrer technischen Anwendungswelt (…))

**Final remarks**

The quote from E. A. Burtt comes from his book *The Metaphysical Foundation of Modern Science*: p. 238.

Like all institutions imbued with the will to survive, they knew how to weather the "dissolution" of their foundations

is a quote by Peter Sloterdijk from *Critique of Cynical Reason*, p. 25.

Where is the soul in the empirical science of psychology and the mind in the cognitive sciences of artificial intelligence?

writes Odo Marquard in *Individuum und Gewaltenteilung* (Individual and separation of powers), p. 127.

The sentence:

We are not afraid of the abyss, we build a bridge over it.

comes from Bolz, Norbert: *Der alte weiße Mann* (The old white man).

The world of reflection and the lifeworld must not become unrelated to each other writes Odo Marquard in *Zukunft braucht Herkunft* (Future needs Heritage), page 21.

The quote

Have the courage to use your own understanding

comes from Immanuel Kant from *Answering the Question: What Is Enlightenment?*

The quotation from Siddhartha Gautama, slightly modified by me, can be found in: Kalama Sutta: *The Buddha's Charter of Free Inquiry.*

Everything that regulates our coexistence should be worked out in democratic processes...: Omri Boehm rightly points out in his wise book *Radikaler Universalismus* (Radical universalism) that democracy must not lead to a tyranny of the majority.

# Bibliography


Berkeley, George: *A Treatise Concerning the Principles of Human Knowledge,* Projekt Gutenberg, 2003

Boehm, Omri: *Radikaler Universalismus* (Radical universalism), Ullstein, 2022

Bolz, Norbert: *Der alte weiße Mann* (The old white man), LMV, 2023

Brodbeck, Karl-Heinz: *Der Zirkel des Wissens* (The circle of knowledge), Shaker Verlag, 2002

Brodbeck, Karl-Heinz: *Handeln als metaphysischer Horizont* (Action as a metaphysical horizon), K.-H. Brodbeck, © 2001

Brodbeck, Karl-Heinz: *Die fragwürdigen Grundlagen der Ökonomie* (The questionable foundations of economics), WBG, 6th ed., 2013

Burtt, E. A.: *The Metaphysical Foundations of Modern Science,* Dover Publications, 2003

Einstein, Albert and Infeld, Leopold: *The Evolution of Physics,* Cambridge University Press, 1938

Einstein, Albert: *The Principle of Relativity,* Library of Alexandria, 2022

Galileo Galilei: *Il Saggiatore,* Rome, 1623

Garfield, Jay L.: *The Fundamental Wisdom of the Middle Way, Nagarjuna's Mulamadhyamakakarika,* Oxford University Press, 1995

Hoffmann, Martin: *A Local Realistic Explanation of EPR Correlations,* Foundations of Physics, Vol. 20, No. 8, 1990

Hoffmann, Martin: *Fernkorrelationen in der Quantentheorie: Eine neue Interpretation* (Long-distance correlations in quantum theory: A new interpretation), Doctoral thesis, University of Tübingen, 1988

Hume, David: *An Inquiry into the Human Mind*, e-artnow, 2019

Kalama Sutta: *The Buddha's Charter of Free Inquiry*, translated from the Pali by Soma Thera. Access to Insight (BCBS Edition), 2013

Kant, Immanuel: *The Critique of Pure Reason*, Projekt Gutenberg, 2003

Kant, Immanuel: *An Answer to the Question: What Is Enlightenment?*, Hackett Publishing, 1992

Koestler, Arthur: *The Sleepwalkers*, Penguin, 2017

Marquard, Odo: *Individuum und Gewaltenteilung* (Individual and separation of powers), Reclam, 2004

Marquard, Odo: *Zukunft braucht Herkunft* (Future needs heritage), Reclam, 2020

Nicholson, Daniel J.: *Is the Cell Really a Machine*?, Journal of Theoretical Biology 477 (2019) 108–126

Ortega y Gasset, José: *The Revolt of the Masses*, W. W. Norton & Company, Inc., 1932

Pirsig, Robert M.: *Zen and the Art of Motorcycle Maintenance*, Mariner Books, 2009

Rosenfield, Israel: *The Strange, Familiar and Forgotten*, Knopf, 1992

Sloterdijk, Peter: *Critique of Cynical Reason*, University of Minnesota Press, 1988

Whitehead, Alfred North: *Science and the Modern World*, Z & L Barnes Publishing, 2022

Zeno of Elea: en.wikipedia.org/wiki/Zeno_of_Elea

# Acknowledgements

I thank Otto Rössler and Hans-Jürgen Machulla, who always believed that I could think for myself.

I am grateful to Gerold Baier, Mites van Oepen, and Marwin Wandrey for their valuable advice on writing the manuscript.

And finally, I thank Daniel Ross for turning my manuscript into a readable text.